Th

Studies in Biology no. 101

Sulphur in Biology

John W. Anderson
Ph.D.
Reader, School of Biological Sciences,
La Trobe University, Victoria, Austra

Edward Arnold

First published 1978
by Edward Arnold (Publishers) Limited
41 Bedford Square, London, WC1B 3DQ

British Library Cataloguing in Publication Data

Anderson, John W
Sulphur in biology. – (Institute of Biology. Studies in biology; no. 101; 0537-9024).
1. Sulphur in the body 2. Sulphur – Physiological effect 3. Plants, Effect of sulphur on
I. Title II. Series
574.1'921 QP535.S1

ISBN 0-7131-2711-2

Printed and bound in G
The Camelot Press Ltd,

General Preface to the Series

Because it is no longer possible for one textbook to cover the whole field of biology while remaining sufficiently up to date the Institute of Biology has sponsored this series so that teachers and students can learn about significant developments. The enthusiastic acceptance of 'Studies in Biology' shows that the books are providing authoritative views of biological topics.

The features of the series include the attention given to methods, the selected list of books for further reading and, wherever possible, suggestions for practical work.

Readers' comments will be welcomed by the Education Officer of the Institute.

1978

Institute of Biology
41 Queen's Gate
London SW7 5HU

Preface

Sulphur is one of the major elements essential for life and has recently been the subject of much biological interest. In many parts of the world the concentration of sulphur in the soil is too low for maximum plant productivity and also adversely affects the nutritional quality of plant foodstuffs which are used for human and animal consumption. But man's interest in the effect of sulphur on organisms has been roused by other events too, such as the burning of fossil fuels which releases toxic gaseous sulphur oxides into the atmosphere.

The issues mentioned above raise fundamental biological questions. Why is sulphur essential for organisms? What forms of sulphur satisfy their nutritional requirements? How is sulphur cycled from one organism to another? What are the world's reserves of sulphur? Will the toxic sulphur oxides accumulate in the atmosphere? This booklet aims to establish certain principles pertinent to the role of sulphur in biology and to show that sulphur is in a state of dynamic equilibrium between organisms in much the same way as nitrogen and carbon. This information is relevant to the questions posed above and hopefully it will stimulate an interest in this important element.

I would like to express my gratitude to Dr R. E. Williamson for many helpful comments in preparing this booklet.

Melbourne, 1977

J. W. A.

Contents

1 Requirements of Organisms for Sulphur

Sulphur is one of the elements essential for the life of all organisms – plants, animals and microorganisms. The requirement of an organism for sulphur can be demonstrated by depriving it of its appropriate sulphur source. Under these conditions the organism will either fail to grow, or, if it contains appreciable reserves of sulphur (e.g. seeds of plants), it will exhibit a slow growth rate, loss of mass or deficiency symptoms compared with a control organism supplied with sulphur. For our purposes, some idea of an organism's requirement for sulphur can be gained from the elemental composition of organisms. Sulphur accounts for approximately 0.2–0.7% of the dry mass of most organisms. On a molar basis several other elements are present in substantially higher concentrations (Table 1).

Table 1 Typical values for the elemental composition of some organisms. All values are expressed as atoms of the element relative to one atom of sulphur. Some of the trace elements are not included. Gaps in the table indicate that values were not determined in conjunction with those for the elements shown.

	Organism		
Element	*Saccharomyces cerevisiae* (yeast)	*Zea mays* (maize)	*Homo sapiens* (man)
Hydrogen		1175	306
Carbon		684	191
Oxygen		523	37.5
Nitrogen	47.5	19.6	27.3
Potassium	4.57	4.43	1.14
Calcium	0.08	1.08	4.78
Phosphorus	3.12	1.21	4.12
Magnesium	0.49	1.40	0.27
Sulphur	1.00	1.00	1.00
Chlorine		0.74	0.54
Iron	0.02	0.27	0.009
Sodium	0.52		0.84

1.1 Sulphur as a constituent of proteins

Carbon, hydrogen and oxygen are associated with almost all of the large number of compounds found in biological matter. The two most fundamental compounds essential for life are the nucleic acids, which also contain nitrogen and phosphorus, and the proteins, which contain nitrogen and sulphur. Cells contain a large number of different proteins which fulfil several essential functions. Some proteins, called enzymes, catalyse the multitude of different chemical reactions associated with the generation of cellular energy and its use in cellular maintenance, growth, development, regulation and reproduction. Other proteins do not have catalytic activity but serve a structural function. Typical of this latter group are some of the proteins associated with membranes such as those of nuclei, mitochondria, chloroplasts and the plasma membrane. Another group of proteins act as a reserve of amino acids and energy.

Proteins are made up of amino acids. The structure of two typical amino acids found in proteins can be written as

where R is different for each amino acid. In proteins the —COOH group of one amino acid is linked to the $—NH_2$ group of another. The —CO—NH— bond formed is known as a peptide bond.

Proteins vary in their size. The animal hormone corticotrophin for example contains 39 amino acid units, the enzyme ribonuclease from ox pancreas contains 124, and many more complex enzymes contain more than 1000. Although there are many thousands of proteins in any one organism, all are built up from about twenty different amino acids. Proteins differ only in the proportion of the twenty amino acids and in the sequence in which they are joined together.

Two of the protein amino acids, cysteine and methionine, contain sulphur in the R-side chain:

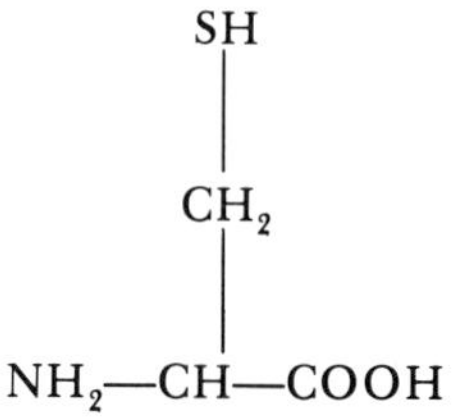

Cysteine

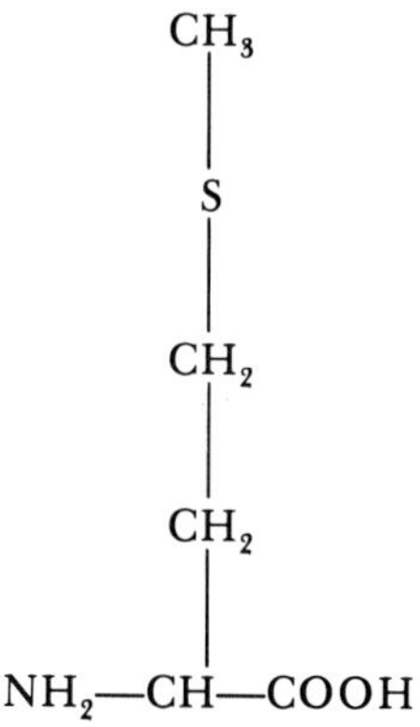

Methionine

If all twenty amino acids were present in a protein in equal molar amounts then the proportion of any single amino acid would be 5% and the cumulative proportion for the two sulphur amino acids would be 10%. For the majority of proteins the proportion of the sulphur amino acids is less than this (typically 3–7%), but the proportions can vary greatly. Keratin, the fibrous protein of hair and wool, contains a very high proportion of cysteine (12%), whereas collagen, a fibrous protein of skin and cartilage, contains no cysteine and only a very small amount of methionine. Nevertheless, it can be said that virtually all proteins contain at least one of the sulphur amino acids. Although sulphur has other essential roles in organisms (see section 1.2), it is the occurrence of sulphur in the protein amino acids cysteine and methionine which constitutes the basic biological importance of this element in all organisms.

Sulphur occurs in cysteine in a —SH group (also known as a thiol or sulphhydryl group). Two aspects of its chemistry are relevant to the role of cysteine in proteins. The —SH group of cysteine, together with the —SH group of another molecule of cysteine, readily oxidizes to form a molecule of cystine. The two molecules are linked by a disulphide bond (—S—S—):

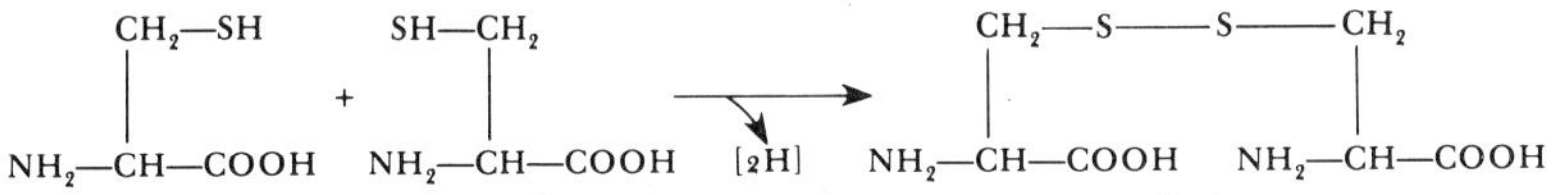

Cysteine *Cysteine* *Cystine*

The same process can occur between two non-adjacent cysteine units within a protein except that in this case the process of oxidation is carefully controlled by the cell. If two cysteine units within a protein are linked together by a —S—S— bond then this will cause the protein to fold (Fig. 1–1). Disulphide linkages are therefore important in determining the configuration of proteins. This in turn is directly relevant to the catalytic or structural properties of the protein.

The other important feature of the chemistry of cysteine is that the —SH group readily reacts with a group of compounds known as —SH group reagents (e.g. Hg^{2+}, Pb^{2+}, *p*-chloromercuribenzoate, *N*-ethylmaleimide). It follows that any —SH groups in the protein which are not oxidized as described above will react with —SH group reagents. It is found that the catalytic activity of a large number of enzymes (perhaps as many as 40%) is destroyed or inhibited by —SH group reagents. This implies that the —SH groups of cysteine are involved in the catalytic function of these enzymes. Incidentally, the reactivity of Hg^{2+} and Pb^{2+} with the —SH groups of proteins largely explains the toxicity of mercury and lead.

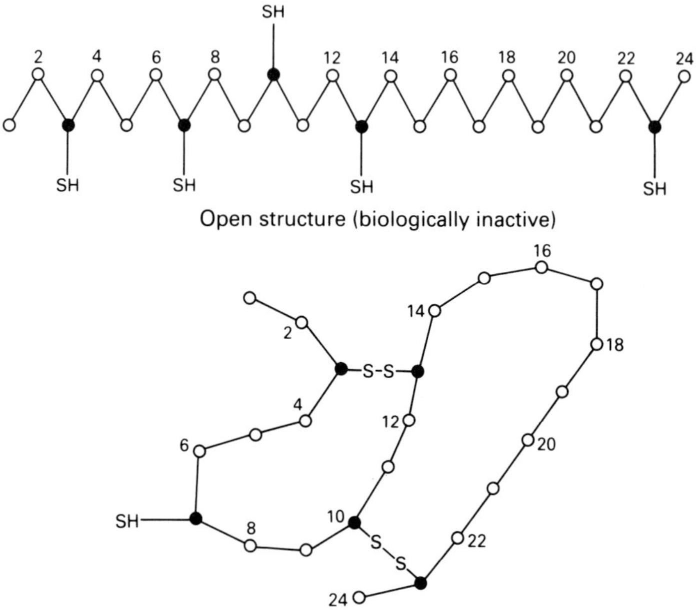

Open structure (biologically inactive)

Folded structure with disulphide bonds (biologically active)

Fig. 1–1 Folding of a hypothetical protein chain by oxidation of the —SH groups of cysteine units in the protein. The precise nature of the folding is determined by the amino acid sequence in the protein. The —SH groups of cysteine at positions 3 and 10 have oxidized with those at positions 13 and 23 respectively to form disulphide bonds (—S—S—) but the —SH group at position 7 has not been oxidized. The unfolded form of the protein is biologically inactive; the folded form is active. If the protein was an enzyme which was completely inhibited by —SH group reagents, we would conclude that the unoxidized —SH group at position 7 was associated with the catalytic function of the enzyme. (●) cysteine units, (o) other amino acid units.

1.2 Other sulphur-containing metabolites

Organisms contain other metabolites, in addition to cysteine and methionine, which contain sulphur. Some of these serve important metabolic roles. Coenzyme A is probably the most important of these. It is a complex molecule whose structure is shown below.

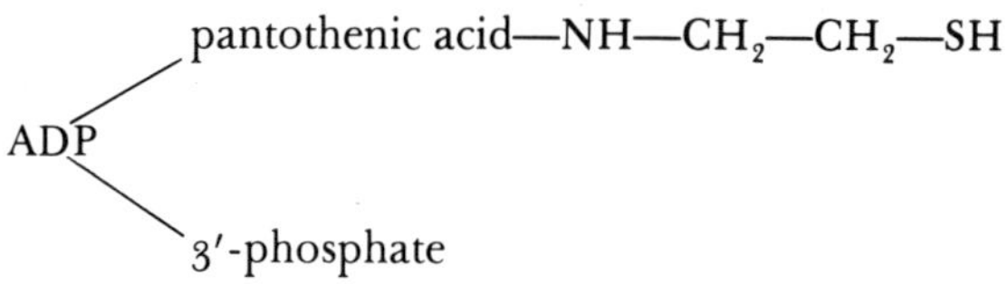

Aminoethane thiol (NH_2—CH_2—CH_2—SH), which can be viewed as decarboxylated cysteine, is linked via its —NH_2 group to pantothenic acid leaving the —SH group free at the terminal end. For this reason coenzyme A is frequently written as CoASH. The biological importance of coenzyme A is that it reacts via its —SH group with carboxylic acids in a series of enzyme-catalysed reactions to form the corresponding thioethers (e.g. acetate is metabolized as acetyl-coenzyme A, CoA—S—CO—CH_3). The metabolic machinery of cells is geared to metabolize many carboxylic acids as their thioethers of coenzyme A. For example, many parts of such basic processes as the oxidation and synthesis of fatty acids, the synthesis of amino acids and the oxidation of certain intermediates of the tricarboxylic acid cycle (carbohydrate metabolism) all involve the formation of the corresponding thioethers of coenzyme A. It is evident that coenzyme A is as intrinsic to metabolism as ATP itself. Since the aminoethane thiol present in the coenzyme A molecule is formed by decarboxylation of cysteine, this is another reason for the importance of this sulphur amino acid in organisms.

The group of compounds known as the ferredoxins (a type of non-haem iron sulphur protein) merit special attention. They are proteins of low molecular weight and contain a higher proportion of cysteine units than usual. The relevance of the ferredoxins to this section rather than section 1.1 relies on the fact that they contain equal numbers of sulphur and iron atoms in addition to the sulphur contained in the cysteine and methionine units in the protein chain. The number of iron and additional sulphur atoms in a ferredoxin molecule depends on the source; the ferredoxin from *Clostridium pasteurianum* (a bacterium) contains eight whereas that from spinach chloroplasts contains two. The exact spatial arrangement of the iron and additional sulphur atoms is not known, though it seems likely that the atoms of iron are linked together via the additional sulphur atoms and that this association is in turn linked to the protein chain via the sulphur atoms of cysteine units in the chain. A typical model for ferredoxin is shown in Fig. 1–2. This configuration confers on the protein a highly negative redox potential, the most negative known for a biological compound; i.e. the reduced form of ferredoxin can, in theory, effect the reduction of oxidized forms of other redox pairs having a more positive redox potential (e.g. reduction of NADP to $NADPH_2$). In plants the importance of ferredoxin lies in the fact that the oxidized form is the recipient of electrons ejected from chlorophyll during photo-oxidation of chlorophyll in the light reactions of photosynthesis (i.e. ferredoxin is reduced). Reduced ferredoxin is probably the eventual source of reducing power for the reduction of CO_2 in the dark reactions of photosynthesis, $NADPH_2$ acting as an intermediate in this process. Reduced ferredoxin also serves as the reducing agent in other cellular processes including nitrogen fixation in

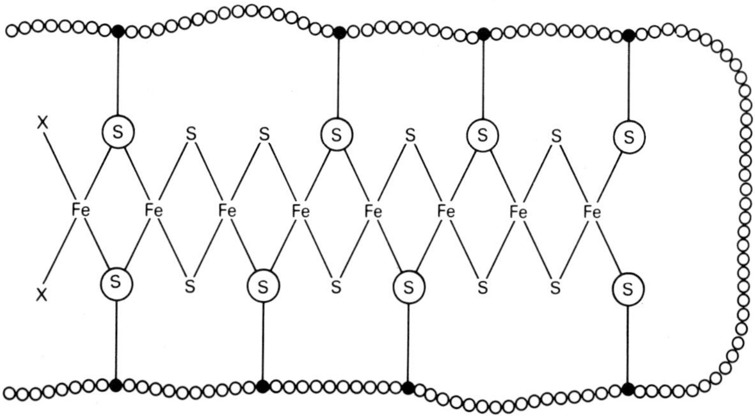

Fig. 1–2 A proposed structure for ferredoxin from the nitrogen-fixing organism *Clostridium pasteurianum*. The outer series of circles enclosing the structure represents the protein chain; cysteine units and other amino acids in the chain are shown by the symbols (●) and (o) respectively. Sulphur atoms enclosed in a circle Ⓢ represent the sulphur atoms of cysteine units. (Redrawn from HUGHES, M. N. (1972). *The Inorganic Chemistry of Biological Processes*. John Wiley and Sons, Inc., New York.)

some nitrogen-fixing organisms. Other iron-sulphur proteins also form part of the electron transport chain of mitochondria.

Further examples illustrating the role of sulphur-containing compounds in metabolic processes could be cited. Some of these are summarized in Table 2.

Table 2 Functions of some sulphur-containing compounds found in organisms.

Compound	*Function*
Cysteine, methionine, co-enzyme A and ferredoxin	See the text.
S-Adenosyl methionine	Methyl group donor for methylation reactions.
Glutathione	Reducing agent, especially important in the reductive detoxification of certain metabolites injurious to cells.
Lipoic acid	Cofactor required in the oxidation of oxo acids in the TCA cycle of aerobic organisms.
Biotin	Cofactor for various carboxylation reactions.
Sulpholipid	Structural material of chloroplast membranes.
Thiamine pyrophosphate	Cofactor for decarboxylation of oxo acids and transketolase reactions.
Formyl methionine	Amino acid of N-terminal end of proteins during synthesis; often cleaved subsequently.

1.3 Environmental sources of sulphur, carbon and nitrogen for organisms

Since all organisms contain sulphur, carbon and nitrogen in organic compounds it follows that at least some organisms must be able to utilize sources of these elements which are freely available in the environment in which they grow. All three elements occur in a variety of forms in geological deposits (e.g. carbon as limestone, $CaCO_3$; sulphur as metal sulphides, FeS, CuS; nitrogen as saltpetre, KNO_3), but these are not available to the majority of organisms because of their geographic localization and their depth below the earth's surface. The sources of carbon, sulphur and nitrogen which are widely available to organisms are located in the top few metres of the earth's crust, the oceans and the gases of the atmosphere. All three elements most frequently occur in these media in oxidized forms: sulphur as sulphate, carbon as CO_2 and nitrogen as nitrate. However, most organic molecules found in cells contain all three elements in reduced forms or, more correctly, low oxidation states (Table 3). This implies that those organisms which can

Table 3 Most common forms of sulphur, carbon and nitrogen in the physical environment used by organisms, and some examples of the forms found in organic compounds in organisms.

Element	*Form found in environment*	*Forms found in organisms*	
Sulphur	SO_4^{2-}	—SH	cysteine
		—S—	methionine
Carbon	CO_2, HCO_3^-	—HCOH—	sugars
		$—CH_2—$	fats
Nitrogen	NO_3^-	$—NH_2$	amino acids
		—NH—	proteins

use the oxidized forms (high oxidation states) present in the environment must possess mechanisms for reducing them. The formation of sugars from CO_2 (reduction) is mainly achieved by photosynthetic organisms. Those organisms which can reduce the sulphur in sulphate to lower oxidation states (e.g. to sulphite and sulphide) are discussed in Chapter 2.

1.4 Bioenergetics of reducing sulphate, carbon dioxide and nitrate

The biological reduction of sulphate, CO_2 and nitrate are all energy-requiring processes. The energy requirements of photosynthesis (reduction) and the energy yield of respiration (oxidation) are probably well known to most students; here they will be briefly compared to the oxidation and reduction of sulphur, carbon and nitrogen to establish a common principle. Considering the oxidation reactions first, glucose, which can be assumed for the purpose of discussion to be the product of photosynthesis, can be oxidized by molecular oxygen in a bomb calorimeter and its change in free energy measured. Heat energy is released in the process implying that the reaction products contain less free energy than the substrates. There has therefore been a loss of free energy (negative ΔG) and the reaction is said to be exergonic. Similarly H_2S and NH_3 (low oxidation state forms of sulphur and nitrogen respectively) can also be oxidized in exergonic reactions:

(1) $C_6H_{12}O_6 + 6O_2 \rightleftharpoons 6CO_2 + 6H_2O$ (ΔG −2870 kJ mol^{-1})
(2) $H_2S + 2O_2 \rightleftharpoons H_2SO_4$ (ΔG −715 kJ mol^{-1})
(3) $NH_3 + 2O_2 \rightleftharpoons HNO_3 + H_2O$ (ΔG −347 kJ mol^{-1})

The magnitude of the ΔG values indicates how far the reactions lie to the right since ΔG is related to the equilibrium constant (K_{eq}) by the equation:

$$\Delta G = -5.7 \log_{10} K_{eq}$$

Thus if ΔG is −5.7 kJ mol^{-1}, K_{eq} is 10; if ΔG is −11.4 kJ mol^{-1}, K_{eq} is 100. Conversely, if ΔG is +5.7 kJ mol^{-1}, K_{eq} is 0.1; and if ΔG is +11.4 kJ mol^{-1}, K_{eq} is 0.01. It follows from this analysis that the substrates in reactions (1) to (3) are almost completely oxidized at chemical equilibrium. In cells, these reactions can serve as sources of energy. Most cells have energy-generating mechanisms based on the aerobic and/or anaerobic oxidation of glucose or related compounds (respiration) but some organisms described in the next chapter use the oxidation of H_2S (and other forms of sulphur with low oxidation states) as energy-generating mechanisms. The same principle applies to the oxidation of forms of nitrogen with low oxidation states.

Now consider reactions (1) to (3) in the opposite direction:

(4) $6CO_2 + 6H_2O \rightleftharpoons C_6H_{12}O_6 + 6O_2$ (ΔG +2870 kJ mol^{-1})
(5) $H_2SO_4 \rightleftharpoons H_2S + 2O_2$ (ΔG +715 kJ mol^{-1})
(6) $HNO_3 + H_2O \rightleftharpoons NH_3 + 2O_2$ (ΔG +347 kJ mol^{-1})

The free energy changes of these reactions are positive and the reactions are said to be endergonic. It follows from the equation relating ΔG and K_{eq} that it is not possible to produce significant amounts of the products in reactions (4) to (6). Yet it is known that some organisms reduce sulphate to H_2S and incorporate it into the sulphur amino acids. It is also known that photosynthetic organisms form sugars and fats from CO_2 whilst some organisms reduce nitrate to NH_3 and incorporate it into amino acids. Clearly, organisms cannot carry out these reductions by the single step reactions shown in reactions (4) to (6) since good yields of product can only be obtained by reactions with a negative ΔG. The ability of organisms to carry out these seemingly impossible reactions lies in their capacity to couple the endergonic reaction (positive ΔG) to exergonic reactions (negative ΔG). Whilst this might be achieved by a series of complex reactions, the ΔG of the overall process is readily calculated by summing the component ΔG values. Using photosynthesis as an example: the endergonic reaction

(7) $6CO_2 + 6H_2O \rightleftharpoons C_6H_{12}O_6 + 6O_2$ (ΔG +2870 kJ mol^{-1})

is coupled to the exergonic reactions

(8) $12NADPH_2 + 6O_2 \rightleftharpoons 12NADP + 12H_2O$ (ΔG −2640 kJ mol^{-1})
(9) $18ATP \rightleftharpoons 18ADP + 18P_i$ (ΔG −552 kJ mol^{-1})

The net ΔG when reactions (7) to (9) are summed is −322 kJ mol^{-1}, thus ensuring that a good yield of glucose is obtained at chemical equilibrium. Glucose has therefore been synthesized at the expense of hydrolysing ATP and oxidizing $NADPH_2$ which in this example are products of the light reactions of photosynthesis. The important point is that metabolic energy has been expended to effect the formation of glucose from CO_2. The same principle is equally true for the reduction of oxidized forms of sulphur and nitrogen to lower oxidation states.

1.5 Biological cycling of sulphur, carbon and nitrogen

The brief treatment of the bioenergetics of sulphur, carbon and nitrogen brings us to the point at which the biological cycles for these elements can be established. Some organisms can, in energy-requiring reactions, use oxidized forms of the elements present in the physical environment to make the organic molecules of life. The energy for these processes can come from the sun, or, in some cases, from the oxidation of other materials present in the environment. Green plants belong to this category and are unique in being able to use sulphate, CO_2 and nitrate as their sources of sulphur, carbon and nitrogen respectively. The reduced forms of the elements, made by these organisms, can in turn be oxidized by a second class of organisms in energy-generating reactions and then used as a source of energy for growth. Animals belong to the latter class. However, it should be noted that many organisms fall into one class with respect to one element but another class with respect to another element. The examples cited in Table 4 illustrate this point. Finally, it should be pointed out that organisms belonging to the first class with respect to a particular element are dependent on the second class for regenerating the oxidized forms of the element for re-use. The two classes of organisms are therefore dependent on each other for the alternate formation of oxidized and reduced forms.

Organisms which can use CO_2 as their sole source of carbon are referred to as autotrophs (*auto*, Gk., self; *trophikos*, Gk., nourishment).

Theoretically this nomenclature could be used for referring to organisms which use sulphate and nitrate as their sources of sulphur and nitrogen for growth (e.g. S-autotrophs and N-autotrophs), but this nomenclature would be very confusing particularly for those organisms which can use sulphate as their sulphur source but cannot use CO_2 as a source of carbon (Table 4). There are other complications too. Some anaerobic bacteria

Table 4 Forms of sulphur, carbon and nitrogen used by some organisms.

Organism	*Sulphur*	*Carbon*	*Nitrogen*
Plants	SO_4^{2-}	CO_2	NO_3^-
Monogastric animals	Organic S	Organic C	Organic N
Ruminant animals	Organic S and SO_4^{2-}	Organic C	Organic N and NH_4^+
Penicillium (a fungus)	SO_4^{2-}	Organic C	NO_3^-
Thiobacillus (a bacterium)	S^{2-}, $S_2O_3^{2}$	CO_2	NO_3^-
Chlorobium (a bacterium)	S^{2-}	CO_2	N_2, NH_4^+

use sulphate as a terminal electron acceptor (thereby effecting its reduction) in a process analogous to the reduction of oxygen by respiration in aerobic organisms; they produce far more sulphide than they require for the synthesis of the sulphur amino acids. Accordingly, in this book, the form of sulphur used by an organism will always be specified.

Organisms which require forms of carbon more complex than CO_2 (i.e. reduced forms of carbon) are referred to as heterotrophs; they must derive their supply of carbon either directly or indirectly from autotrophs. Here again there is an analogy between sulphur and carbon; some organisms are unable to synthesize the sulphur amino acids from any form of inorganic sulphur regardless of the oxidation state. Such organisms are entirely dependent on those that can.

2 Transformations of Inorganic Sulphur by Organisms

In the previous chapter it was established that organisms catalyse two fundamental transformations of sulphur: oxidation, an energy-generating process, and reduction, an energy-requiring process. It was also noted that most organisms capable of reducing oxidized forms of sulphur incorporate the sulphur they reduce into the sulphur amino acids and that these organisms provide a supply of cysteine and methionine for those organisms incapable of using inorganic sulphur. This chapter is concerned with the organisms that catalyse interconversions of inorganic sulphur, the reactions involved, the importance of the reactions in the life of the organisms and how the reactions are possible in relation to the principles of bioenergetics. In addition it will be seen that the growth of the various organisms is restricted to those habitats which provide suitable conditions for the operation of the appropriate biological reactions that they catalyse.

2.1 Oxidation states of sulphur

Sulphur can exist in a large number of oxidation states. Some of these are shown in Table 5. Sulphate is the most oxidized form of sulphur

Table 5 Some inorganic forms of sulphur and the changes in free energy associated with their oxidation to sulphate by molecular oxygen.

Form	*Formula*	*Oxidation state*	*ΔG of oxidation to 1 mol of sulphate* (kJ mol^{-1})
Sulphate	SO_4^{2-}	+6	0
Dithionate	$S_2O_6^{2-}$	+5	−139
Sulphite	SO_3^{2-}	+4	−200
Disulphite	$S_2O_5^{2-}$	+4	
Dithionite	$S_2O_4^{2-}$	+3	−327
Thiosulphate	$S_2O_3^{2-}$	+2	−497
Elemental sulphur	S^0	0	−502
Sulphide	S^{2-}	−2	−715

(oxidation state +6) and therefore can only act as an oxidizing agent. It is one of the most stable forms of sulphur and is the most abundant form of sulphur available in the physical environment for use by organisms.

Under specialized local conditions other forms of sulphur can be present in substantial amounts. Waterlogged (anaerobic) soils, for example, frequently contain sulphur in lower oxidation states (e.g. sulphide and thiosulphate) and they contribute to the unpleasant smell of these soils. However, sulphur in lower oxidation states is not normally present in significant amounts in well-drained soils.

The free energy changes associated with the oxidation of the various forms of sulphur to sulphate by molecular oxygen are shown in Table 5. All these reactions, have a negative ΔG, the value becoming increasingly more negative with decreasing oxidation states. Incidentally, oxidation does not imply that sulphate is formed; oxidation only entails an increase in oxidation state (e.g. thiosulphate to sulphite, sulphide to thiosulphate, etc.).

The sulphur amino acids essential for life contain sulphur in oxidation state -2. The same oxidation state is found in H_2S and the sulphides. It follows therefore that if an organism uses an inorganic sulphur source with an oxidation state greater than -2 for the synthesis of the sulphur amino acids then the sulphur must first be reduced to oxidation state -2. Organisms that use oxidized forms of sulphur carry out the reduction and then subsequently incorporate sulphide into an acceptor molecule to form the sulphur amino acid cysteine. It follows that the input of free energy required to obtain a given yield of sulphide is greatest for those forms of sulphur with the highest oxidation states.

2.2 The major types of inorganic sulphur transformations catalysed by organisms

Organisms catalyse a wide variety of sulphur interconversions involving both oxidation and reduction. Each conversion must comply with the principles of bioenergetics. Clearly it is erroneous to consider, for example, the oxidation of sulphide to sulphate by one organism as the opposite process to the reduction of sulphate to sulphide in another. As in the formation of sugars from CO_2 in photosynthesis and the oxidation of sugars to CO_2 in aerobic respiration, the processes involved in sulphur oxidation and reduction involve completely different biological processes. Indeed, even the biological reduction of sulphate can be achieved by two distinctly different processes. The transformations of inorganic sulphur catalysed by organisms can be classified according to the biological and bioenergetic principles involved; these are summarized in Table 6.

2.3 Assimilatory sulphate reduction

The process of assimilatory sulphate reduction involves the expenditure of metabolic energy to reduce inorganic sulphate (or any

Table 6 Classification of the interconversions of sulphur catalysed by organisms.

Process	*Typical organisms*	*Mechanism of sulphur transformation*
A Reduction of oxidized forms		
I Assimilatory		
(*a*) Photosynthetic	Green plants	Sulphate reduced to sulphide by reduced ferredoxin generated by the light-dependent reactions of the chloroplast electron transport chain. Sulphide used for the synthesis of sulphur amino acids.
(*b*) Aerobic heterotrophic	Most fungi (e.g. yeasts and *Penicillium*) and most aerobic bacteria	As for plants except that the reducing agent ($NADPH_2$) is generated by aerobic oxidation of a heterotrophic substrate (e.g. glucose).
II Dissimilatory	Members of the bacterial genera *Desulfovibrio* and *Desulfotomaculum*	Anaerobically oxidize heterotrophic substrates and use sulphate as terminal electron acceptor causing reduced forms of sulphur to accumulate in the growing medium.
B Oxidation of reduced forms		
I Chemoautotrophic	Members of the bacterial genus *Thiobacillus* and probably several other genera	Aerobically oxidize reduced forms of sulphur to higher oxidation states in exergonic reactions which serve as the source of energy for the assimilation of CO_2.
II Photoautotrophic	Members of the bacterial genera *Chlorobium* and *Chromatium*	H_2S is used as electron donor for the light reactions in a manner analogous to H_2O in green plants causing elemental sulphur to accumulate. Organisms strictly anaerobic.
III Heterotrophic		
(*a*) Sulphur turnover	Animals	Cysteine oxidized to sulphate as part of normal amino acid and protein turnover. Not an important energy source. Sulphate excreted in urine.
(*b*) Desulphurylation	Most anaerobic bacteria	Removal of H_2S from organic sulphur compounds to produce a simple carbon substrate which is used as an energy source. H_2S not necessarily oxidized.

form of inorganic sulphur more oxidized than oxidation state −2) to sulphide which is then incorporated into an acceptor molecule (usually O-acetylserine) to form the sulphur amino acid cysteine. Green plants, most fungi and most aerobic bacteria catalyse assimilatory sulphate reduction and account for almost all of the biological synthesis of the sulphur amino acids. Most of these organisms are also capable of assimilatory nitrate reduction. Indeed there are several analogies between the two processes which can be represented as follows:

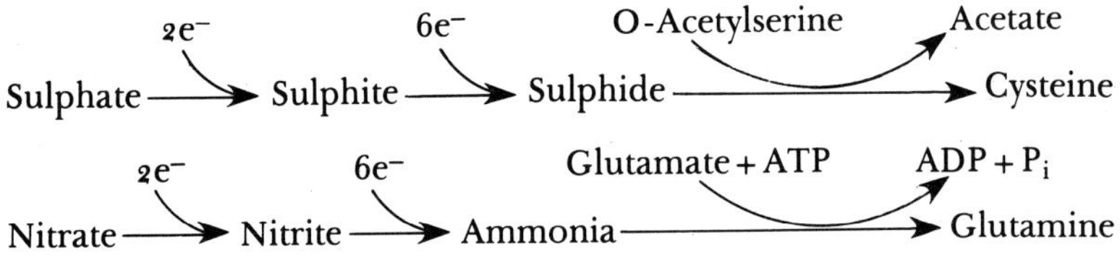

Organisms differ in the reducing agents they use to effect the reduction of sulphate to sulphide. In plants it is believed to be reduced ferredoxin which is generated by the light-dependent photosynthetic electron transport chain of chloroplasts. In fungi and bacteria, it is probably $NADPH_2$ which is generated by the oxidation of reduced carbon substrates (e.g. glucose). Plants and heterotrophic organisms also appear to differ in some detailed mechanistic aspects of the sulphate reduction process. Because of the importance of cysteine and methionine in plant foodstuffs with respect to the nutrition of man and his domestic animals, the synthesis of these compounds from inorganic sulphate is described in more detail in section 3.2.

2.4 Dissimilatory sulphate reduction

Organisms which catalyse assimilatory sulphate reduction synthesize sulphide principally for use in the formation of the sulphur amino acids; they do not release significant amounts of sulphide into the environment nor accumulate sulphide within their cells. Bacteria belonging to the genera *Desulfovibrio* and *Desulfotomaculum*, however, consume relatively large amounts of oxidized forms of sulphur and reduce it almost quantitatively to lower oxidation states. The products either accumulate in the cells or are released into the environment; only a small proportion of the sulphur reduced is used for the synthesis of the sulphur amino acids.

These bacteria, under strictly anaerobic conditions, partially oxidize simple substrates (e.g. succinate or hydrogen) and use the oxidized forms of sulphur (e.g. sulphate) as their terminal electron acceptor thereby producing reduced forms of sulphur (e.g. sulphide). This process is known as dissimilatory sulphate reduction or, alternatively, respiratory sulphate reduction. The principle can be more easily understood by making an analogy with the partial oxidation of succinate by an aerobic organism using oxygen as the terminal electron acceptor. Both organisms catalyse succinate oxidation:

$$\text{Succinate} + \text{X} \longrightarrow \text{Fumarate} + \text{XH}_2$$

Where X is the intermediate electron acceptor (e.g. NAD). XH_2 is then oxidized by the electron transport chain which differs in the two types of organism:

Aerobes	$2XH_2 + O_2$	$\longrightarrow$	$2X + 2H_2O$
Dissimilatory sulphate reducers	$4XH_2 + SO_4^{2-}$	$\longrightarrow$	$4X + S^{2-} + 4H_2O$

Since the oxidation of XH_2 by oxygen or sulphate is an exergonic process this permits the synthesis of ATP during the passage of electrons down the appropriate electron transport chains. The production of sulphide by organisms catalysing dissimilatory sulphate reduction is therefore a consequence of their respiratory mechanism and explains why these organisms produce large amounts of sulphide and/or other reduced forms of sulphur which accumulate in the growing medium.

Dissimilatory sulphate reduction occurs only in the two bacterial genera *Desulfovibrio* and *Desulfotomaculum*. A description and classification of these organisms can be found in the references marked with an asterisk on page 60 of this booklet. The morphology of *Desulfovibrio africanus* is shown in Fig. 2–1a. It follows from the principle of dissimilatory sulphate reduction that these organisms are restricted to anaerobic habitats containing inorganic sulphate (or some fairly oxidized form of sulphur) and a suitable energy source (e.g. a reduced carbon substrate or hydrogen). These conditions are found in waterlogged soils and muds, sewage pipes, drains and pits, stagnant ponds and the rumen of ruminant animals. Although the amount of sulphate consumed by dissimilatory sulphate-reducing bacteria on a global basis is probably fairly insignificant in comparison with that utilized by assimilatory organisms, the dissimilatory organisms are nevertheless of considerable biological and economic importance. For example, one species of *Desulfovibrio* (*D. ruminis*), present in the rumen of ruminant animals, plays an important part in the sulphur nutrition of these animals (see section 3.6), the main source of animal protein for human consumption.

The dissimilatory bacteria are probably better known, however, for the unpleasant odours that they generate and for their participation in the corrosion of buried iron pipes. Various salts in the soil which come in contact with exposed parts of the pipe induce an electrochemical potential which causes the iron to dissolve. This process is accompanied by the evolution of hydrogen gas:

$$(1)\quad 4Fe + 8H_2O \longrightarrow 4Fe^{2+} + 8OH^- + 8H \qquad \text{(polarization)}$$

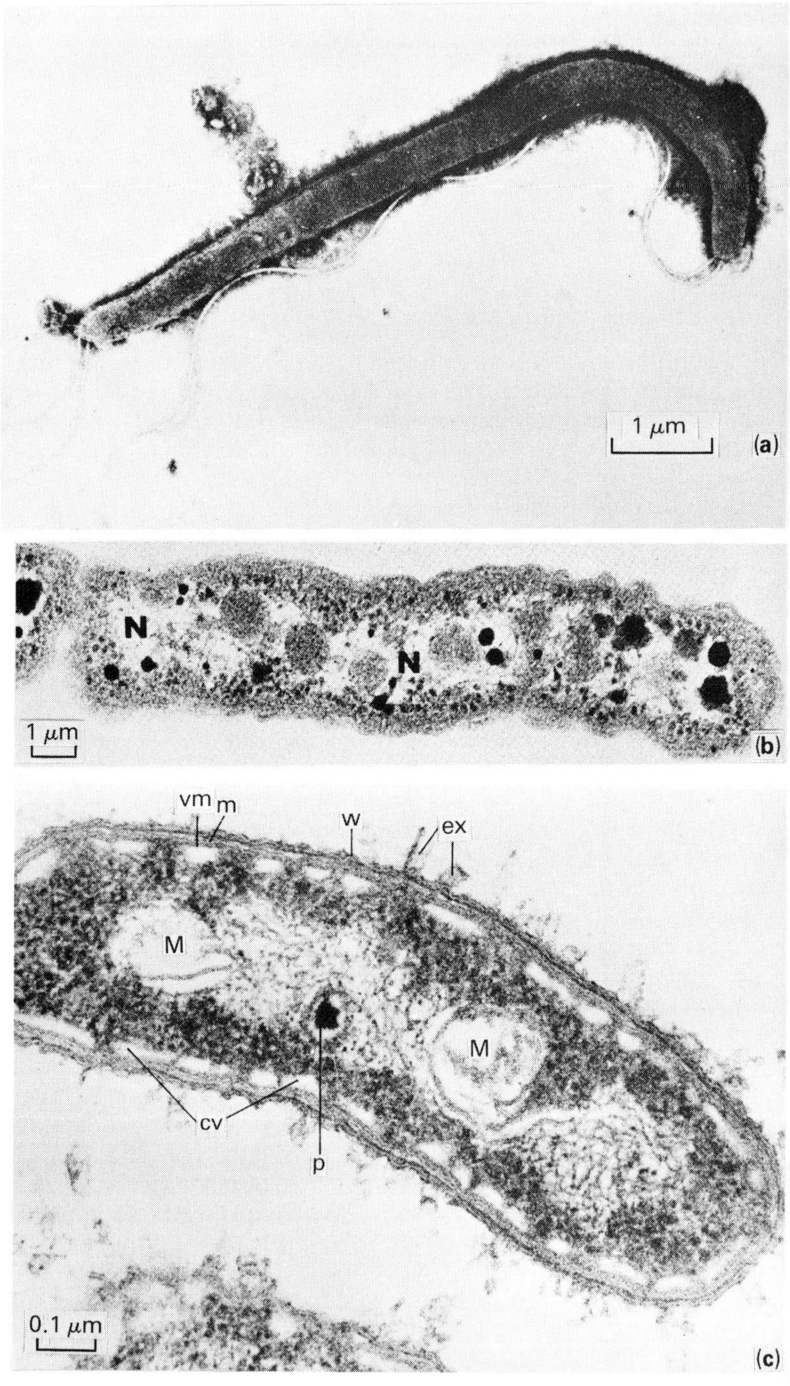
1 µm
(a)
N
N
1 µm
(b)
vm
m
w
ex
M
M
cv
p
0.1 µm
(c)

If the hydrogen is not removed from the surface of the pipe an opposing hydrogen electrode is formed and dissolution of the iron will cease. The pipe is then said to be polarized. However, dissimilatory sulphate-reducing bacteria utilize the hydrogen as an energy source for growth thereby depolarizing the pipe.

(2) $SO_4^{2-} + 8H \longrightarrow S^{2-} + 4H_2O$ (bacterial depolarization)

This will induce further dissolution of iron (reaction (1)). The net products of reactions (1) and (2) react chemically to form the corrosion products FeS and $Fe(OH)_2$:

$$4Fe^{2+} + 8OH^- + S^{2-} \longrightarrow FeS + Fe(OH)_2 + 2OH^-$$

On the beneficial side it is suspected that the dissimilatory sulphate-reducing bacteria have played an important part in the formation of metal sulphide ores (e.g. FeS, CuS, ZnS) and, in conjunction with the photosynthetic sulphur-oxidizing bacteria, the formation of elemental sulphur deposits (see section 4.4). These materials are important sources of metals and sulphur for man's industrial requirements.

2.5 Oxidation of sulphur by chemoautotrophic bacteria

As discussed in section 2.1, inorganic sulphur in low oxidation states is oxidized by molecular oxygen to higher oxidation states in exergonic reactions. The sulphur-oxidizing chemoautotrophic bacteria use these reactions as their primary energy-generating mechanism to provide the free energy required for the synthesis of organic compounds from CO_2. It appears that c-type cytochromes act as intermediate electron acceptors during oxidation of suitable substrates such as sulphide, sulphur (S^0) and thiosulphate. Subsequent electron flow via the bacterial electron transport chain to oxygen then leads to ATP synthesis in a manner analogous to oxidative phosphorylation in the mitochondria of higher organisms. Alternatively, the c-type cytochrome(s) can lose electrons to NADP. The $NADPH_2$ and ATP formed are used for the assimilation of CO_2 in the Calvin cycle in much the same way as in green plants. These

Fig. 2–1 The morphology and fine structure of some bacteria which catalyse transformations of inorganic sulphur. (**a**) Morphology of *Desulfovibrio africanus* showing its curved-rod shaped structure and flagellae attached to each of the two poles of the cell. (**b**) Section of *Thiobocillus neapolitanus*, showing the surrounding gram-negative cell wall and nucleus (N). The dark bodies are granules. (**c**) Longitudinal section of *Chlorobium thiosulfatophilum* showing vesicles (cv), membranes (m and vm), cell wall (w). (**a**) from JONES, H. E. (1971), *Arch. Mikrobiol.*, **80**, 78–86; (**b**) from HOLT, S., SHIVELY, J. M. and GREENAWALT, J. W. (1974). Reproduced by permission of the National Research Council of Canada from the *Can. J. Microbiol.*, **20**, 1347–51. (**c**) from COHEN-BAZIRE, G., PFENNIG, N. and KUNISAWA, R. (1964). *J. Cell Biol.*, **22**, 207–12.

processes are shown diagrammatically in Fig. 2–2. With respect to CO_2 incorporation, plants and the sulphur-oxidizing chemoautotrophic bacteria differ only in the mechanisms used to generate ATP and $NADPH_2$.

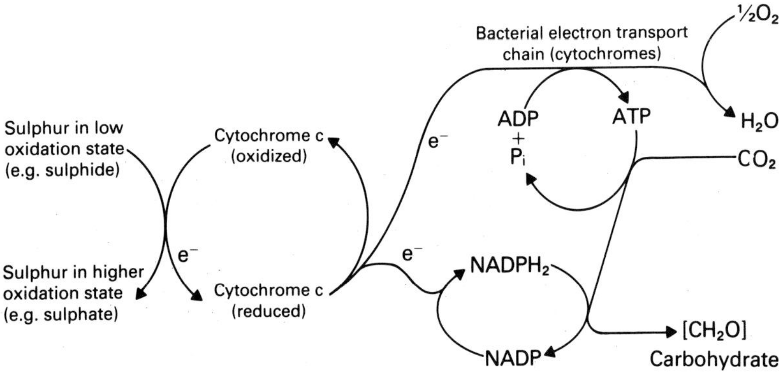

Fig. 2–2 Summary of the component processes of CO_2 assimilation in the sulphur-oxidizing chemoautotroph *Thiobacillus.*

Members of the bacterial genus *Thiobacillus* are sulphur-oxidizing chemoautotrophs. A brief description of these curved or rod-shaped organisms can be found in the references marked with an asterisk on p. 60. The morphology of *Thiobacillus neapolitanus* is illustrated in Fig. 2–1b. *Thiobacillus* species differ somewhat in the forms of sulphur they will oxidize but, generally speaking, most species oxidize sulphide, elemental sulphur and thiosulphate to sulphate. Members of the bacterial genera *Thiovulum* and *Thiospirillopsis* are also sulphur-oxidizing chemoautotrophs. Many other bacteria are also known to oxidize reduced forms of sulphur but in most cases it has not been established that they use CO_2 as a carbon source.

The requirement of sulphur-oxidizing chemoautotrophic bacteria for CO_2, oxygen and an inorganic sulphur source of low oxidation state determines the habitats in which they grow. Since most of the reduced forms of sulphur required by these organisms are produced by the action of anaerobic bacteria (e.g. dissimilatory sulphate reduction) it follows that the sulphur-oxidizing chemoautotrophs tend to occur at the interface between anaerobic conditions (source of sulphide) and aerobic conditions (source of oxygen). They are therefore frequently active near the surface of stagnant ponds and canals, waterlogged soils, muds and marshes. However, the occurrence of reduced forms of sulphur is not restricted to anaerobic habitats. Many microorganisms for instance, produce reduced forms of sulphur during the decomposition of plant, animal and microbial matter in well-drained (aerobic) soils. Sulphur-oxidizing autotrophs are therefore commonly active in these soils and their activity plays an important part in maintaining a supply of sulphate for the growth of plants.

The sulphur-oxidizing chemoautotrophs are well known for the economic damage that they cause. They almost quantitatively convert reduced forms of sulphur to sulphuric acid which is then excreted into the growing medium. In particular *Thiobacillus thio-oxidans* and *Th. concretivorus* are so tolerant of the sulphuric acid they produce that they can decrease the pH of the medium in which they grow to pH 0. Concrete, which is essentially a stone aggregate bonded together by calcium carbonate, is especially sensitive to attack by the sulphuric acid produced by the action of *Thiobacillus*; the acid reacts with the calcium carbonate causing the concrete to 'rot'. Mortar, used in the construction of brick and stone buildings, and certain building stones (e.g. limestone) are damaged in a similar way. Corrosion of concrete is most frequently experienced in sewers. This problem is most severe when reduced forms of sulphur, generated by the action of organisms under the anaerobic conditions prevailing in the sewer, are mixed with oxygen (e.g. surge tanks, holding ponds, etc.) thus providing conditions suitable for the growth of *Thiobacillus*.

2.6 Oxidation of sulphur by photoautotrophic bacteria

All photosynthetic organisms contain a pigment which is photo-oxidized by light. The electrons ejected from the pigment during photo-oxidation are trapped by a primary electron acceptor which in turn donates electrons to other acceptor molecules in exergonic reactions which are coupled to the synthesis of ATP. The final acceptor molecules of the electrons ejected from the pigment are usually ferrodoxin and/or NADP. The reduced forms of these compounds, together with ATP, are in turn used for the assimilation of carbon dioxide, a process common to all photosynthetic organisms:

$$\text{Pigment} \xrightarrow{h\nu} \text{Pigment}^+ + e^-$$

$$\text{Primary acceptor} + e^- \longrightarrow \text{Reduced primary acceptor}$$

$$\text{Reduced primary acceptor} + \text{ADP} + \text{P}_i + \text{NADP} \longrightarrow \text{NADPH}_2 + \text{Oxidized primary acceptor} + \text{ATP}$$

$$\text{ATP} + \text{NADPH}_2 + \text{CO}_2 \longrightarrow [\text{CH}_2\text{O}] + \text{ADP} + \text{P}_i + \text{NADP}$$

This series of reactions, however, does not make for a self-sustaining system unless a mechanism exists for returning the pigment to its unoxidized state. In green plants, this is achieved by withdrawing electrons from water:

$$2\,\text{Pigment}^+ + \text{H}_2\text{O} \longrightarrow 2\,\text{Pigment} + 2\text{H}^+ + \tfrac{1}{2}\text{O}_2$$

Some lower photosynthetic organisms, however, use a variety of other compounds as electron donors to effect the reduction of oxidized pigment. The organisms that concern us here are the green bacteria belonging to the family Chlorobacteriaceae (e.g. *Chlorobium*, Fig. 2–1c) and the purple and red bacteria of the family Thiorhodaceae (e.g. *Chromatium*). These photosynthetic bacteria use H_2S as their source of electrons to return the oxidized pigment (bacteriochlorophyll) to its reduced state, resulting in the formation of elemental sulphur:

$$2\,\text{Pigment}^+ + H_2S \longrightarrow 2\,\text{Pigment} + S^0 + 2H^+$$

The sulphur formed in this process is produced in substantial amounts and is accumulated within the bacterium (e.g. *Chromatium*) or excreted as granules into the growing medium (e.g. *Chlorobium*). It follows that the photosynthetic sulphur-oxidizing bacteria require light, CO_2, H_2S and anaerobic conditions for their growth. Thus these organisms are most commonly found close to the surface of muds and stagnant bodies of water where they receive light but sufficiently deep in the profile for them to be in an anaerobic environment.

Whilst the basic principles of photosynthesis in green plants and photosynthetic bacteria are similar there are some important mechanistic differences. The most important of these is that green plants are aerobic organisms possessing two light reactions whereas the photosynthetic bacteria are anaerobic and have only one light reaction (Fig. 2–3). The light reaction of the photosynthetic bacteria, however, has some similarities to one of the light reactions (photosystem I) of plants. This is of some interest to the evolution of photosynthetic organisms. The present-day photosynthetic bacteria are presumed to be descended from and similar to previous photosynthetic organisms which evolved at a time when most of the biologically important elements occurred in reduced oxidation states (e.g. N as NH_3, S as H_2S, O as H_2O, H as H_2). The changing of the earth's atmosphere to more oxidizing conditions led to a gradual decrease in the availability of H_2S, etc. This placed at a selective advantage any organisms able to use more readily available electron donors. The most abundant of these was water, a molecule, however, with a more positive redox potential than the pigment system already existing in the photosynthetic bacteria (Fig. 2–3). The utilization of water was achieved by the green plants through the evolution of a second light reaction (photosystem II) generating an intermediate electron donor to supply the first pigment. The selective advantage of using the ubiquitous water is amply illustrated by considering the evolutionary success of the green plants relative to organisms using other electron donors.

The fine structure of the photosynthetic apparatus is consistent with this view. The green sulphur-oxidizing bacteria (e.g. *Chlorobium*) have the simplest structure in which the pigment (bacteriochlorophyll) is contained in vesicles (Fig. 2–1c). The purple sulphur-oxidizing bacteria (e.g. *Chromatium*) contain bacteriochlorophyll structurally attached to

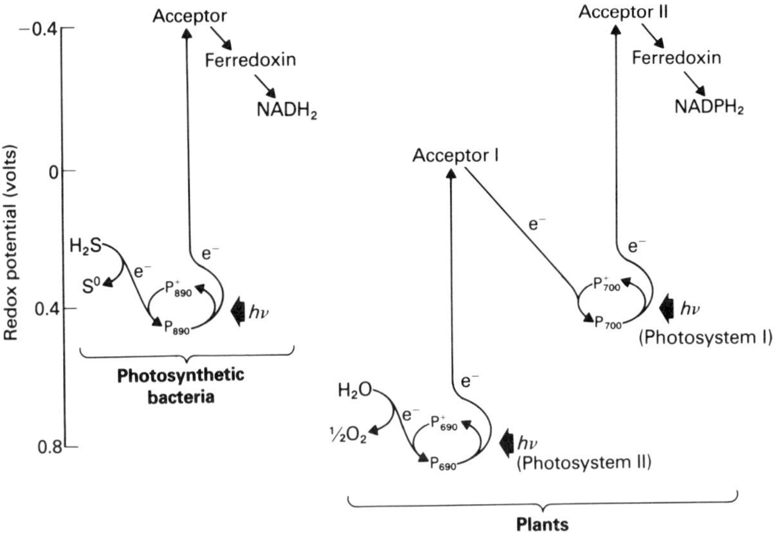

Fig. 2–3 The light reactions of green plants and sulphur-oxidizing photosynthetic bacteria. The symbols P_{890}, P_{690} and P_{700} represent the photo-oxidizable forms of bacteriochlorophyll and chlorophyll present in bacteria and plants respectively.

membranes. The membranes of the purple non-sulphur bacteria (e.g. *Rhodopseudomonas*) also contain chlorophyll. The membranes are arranged in stacks and are therefore considered to be structurally more advanced. Increasing degrees of structural organization and biochemical complexity are progressively found in the prokaryotic blue-green algae and in the chloroplasts of eukaryotic plants.

Today, the sulphur-oxidizing photosynthetic bacteria are relatively unimportant in the amount of sulphur they oxidize and the amount of CO_2 they assimilate compared to other organisms. It is very likely, however, that past generations of these organisms have been main contributors to the formation of the deposists of elemental sulphur which form an important part of the world's sulphur reserves (see section 4.4).

2.7 Formation of inorganic sulphur from organic compounds by heterotrophic organisms

Whereas the processes of sulphur reduction by dissimilatory bacteria and sulphur oxidation by autotrophic bacteria lead to the formation of inorganic forms of sulphur which are returned to the environment, the formation of organic sulphur compounds by organisms (especially those catalysing assimilatory sulphate reduction) does not. It is therefore apparent that continued incorporation of inorganic sulphur into organic sulphur compounds would soon deplete the supply of sulphur available unless some mechanism was available for converting organic sulphur back to inorganic forms. Most aerobic heterotrophic organisms catalyse the oxidation of organic sulphur compounds to inorganic sulphur (usually sulphate) though this process is rarely used as an important energy-generating mechanism – it is more the consequence of maintaining a steady state level of sulphur amino acids within the organism. Animals provide a good illustration of this point. Neglecting growth, a mature animal needs a continual supply of sulphur amino acids (see section 3.3) for the synthesis of protein which forms part of the constant turnover of protein in all organisms. Excess sulphur amino acids (e.g. cysteine), must therefore be excreted. A typical mechanism in animals is as follows:

$$\text{Cysteine} + O_2 \longrightarrow \text{Cysteine sulphinate}$$
$$\text{Cysteine sulphinate} + \text{NAD} + H_2O \longrightarrow \text{Sulphinyl pyruvate} + \text{NADH}_2 + NH_3$$
$$\text{Sulphinyl pyruvate} \longrightarrow \text{Pyruvate} + \text{Sulphite}$$
$$\text{Sulphite} + \tfrac{1}{2}O_2 \longrightarrow \text{Sulphate}$$

$$\text{Sum: Cysteine} + 1\tfrac{1}{2}O_2 + \text{NAD} \longrightarrow \text{Pyruvate} + \text{Sulphate} + NH_3 + \text{NADH}_2$$

The sulphate formed in this process is excreted in the urine. Pyruvate and $NADH_2$ are used as sources of energy for the formation of ATP.

Another example of the conversion of organic sulphur to inorganic sulphur is given by various anaerobic bacteria which use protein as a carbon source. These organisms, however, do not effect the oxidation of sulphur; they simply hydrolyse protein into its component amino acids and eliminate sulphur and nitrogen so that they are left with a carbon source which they then oxidize for the production of energy. Sulphur and nitrogen are therefore released as inorganic forms in the same oxidation states as they occur in the amino acids: sulphur is released as H_2S and nitrogen as NH_3. These products largely account for the stench of putrefying protein. The release of H_2S is known as desulphurylation.

3 Sulphur Nutrition and Metabolism of Plants and Animals

Animals, including man, are unable to make the sulphur amino acids, cysteine and methionine from inorganic sulphur. It is therefore essential that these amino acids are included in the diet of animals and in the correct proportions relative to the other essential amino acids. Green plants, which can synthesize the sulphur amino acids from inorganic sulphate, are the direct or indirect source of these compounds for most animals. Ruminant animals, however, are a special case. This chapter is concerned with the forms of sulphur required by plants, monogastric and ruminant animals (i.e. their nutrition) and how these organisms utilize their respective sulphur sources (i.e. their metabolism).

3.1 Sulphur nutrition of plants

Plants synthesize cysteine and methionine from inorganic sulphate which they obtain from soil via their root system. In a biological context, plants are the greatest consumers of sulphur from the physical environment and the most important producers of sulphur amino acids. It has been estimated that the annual production of plant material (measured as dry mass) is approximately 230×10^9 tonnes. Since dry plant material contains approximately 0.2% sulphur this means that the annual consumption of inorganic sulphur by plants is approximately 0.46×10^9 tonnes. Cereal crops typically consume approximately 20 kg sulphur ha^{-1} $year^{-1}$ while cabbages and onions (which contain high concentrations of so-called secondary plant products rich in sulphur) remove as much as 70 kg sulphur ha^{-1} $year^{-1}$.

The sulphur requirements of plants can be readily demonstrated by growing them in nutrient solutions containing all the elements required for plant growth except sulphur. In the absence of sulphur, plants simply fail to grow. If 1 mM sulphate is added to the nutrient solution the plants grow normally. At the intermediate concentrations the growth is dependent on the concentration of sulphate (Fig. 3–1). If the concentration of sulphate is not sufficient to promote optimum growth the plant will, depending on the species, show various symptoms characteristic of sulphur deficiency. In general, sulphur deficient plants have pale-green leaves, often with red or purple tints; these symptoms are usually most pronounced in the youngest leaves near the apex.

The normal origin of sulphur for plant growth is soil. The relative proportions of the various forms of sulphur in the soil can vary enormously depending on the physical and chemical properties of the

Fig. 3–1 Effect of concentration of inorganic sulphate on the growth of tomato plants (*Lycopersicum esculentum*) raised in water cultures containing all the other nutrient salts required for plant growth.

soil, seasonal conditions, the vegetation that it supports, and whether fertilizers containing sulphur have been supplied. Generally speaking, only 7% of the total sulphur in the top 25 cm is available to plants at any one time in most of the well-drained soils used for agricultural purposes (see section 4.8). Approximately 60–70% of the total sulphur is permanently unavailable. The remainder is mostly associated with organic matter in the soil. The sulphur in this fraction, however, is slowly oxidized to inorganic sulphate by various biological and physical processes ensuring a continual supply of sulphate for plant growth. In general, plants grown in soils containing total sulphur content greater than 0.5 mg g^{-1} dry soil are unlikely to respond to additional sulphur.

3.2 Synthesis of cysteine and methionine by plants

Plants expend metabolic energy in taking up sulphate from soil. The uptake of sulphate is accomplished by a series of specific carrier proteins located in the plasma membrane. The uptake, which occurs against a concentration or energy gradient, is coupled to the hydrolysis of ATP, the latter providing the energy required to effect the process. Collectively these components form the sulphate uptake mechanism which is specific for sulphate; other anions, such as phosphate and nitrate, are not taken up by this mechanism. Some sulphate is reduced in root cells but most of it is transported in the xylem to the leaves where it enters the chloroplasts. CO_2 fixation is not the only process in chloroplasts which is driven by light-generated ATP and $NADPH_2$. We now know that chloroplasts catalyse the assimilation of inorganic nitrogen in light-coupled reactions, and also catalyse the assimilation of inorganic sulphate. On entering the chloroplast, sulphate reacts with ATP to form adenosine sulphatophosphate (APS) and pyrophosphate (PP_i). APS can be viewed simply as AMP with sulphate tacked onto the phosphate group:

$$\underset{\text{(ATP)}}{\text{Adenosine–P–P–P}} + \text{Sulphate} \longrightarrow \underset{\text{(APS)}}{\text{Adenosine–P–sulphate}} + PP_i$$

The APS formed in this reaction is then reduced to sulphide in a complex series of reactions in which reduced ferredoxin (Fe^{++}) generated by the light reactions, serves as the reductant:

$$\text{APS} + \text{Ferredoxin}_{\text{(reduced)}} \longrightarrow \text{AMP} + S^{2-} + \text{Ferredoxin}_{\text{(oxidized)}}$$

Finally sulphide is incorporated into O-acetylserine to form cysteine. O-acetylserine can be viewed as a high energy compound in the same sense as ATP; indeed, ATP is consumed in its formation, the free energy of hydrolysis of ATP being conserved first in the formation of acetyl-CoA and subsequently in O-acetylserine:

ATP + CoASH + Acetate ⟶ Acetyl-CoA + AMP + PP_i
Acetyl-CoA + Serine ⟶ O-acetylserine + CoASH
O-acetylserine + S^{2-} ⟶ Cysteine + Acetate

The high free energy of hydrolysis of O-acetylserine, which occurs during assimilation of sulphide, ensures that the last reaction has a large negative ΔG thereby ensuring a good yield of cysteine at chemical equilibrium.

The light reactions supply energy (ATP) and reducing equivalents (ferredoxin) for these reductive biosyntheses as they do for the biosynthesis involving CO_2. Indeed, by the conventional restriction of the term 'photosynthesis' to carbon assimilation alone, the fact is obscured that CO_2 is only one – albeit the most important – inorganic substrate reduced by the chloroplast.

Cysteine in turn is the starting point for the synthesis of most other sulphur-containing compounds in plants including coenzyme-A, methionine, S-adenosyl methionine and glutathione. The biosynthesis of methionine is of considerable interest as it is one of the essential amino acids required by monogastric animals; deficiencies of methionine and cysteine in certain foodstuffs affects the nutrition of many people in the developing nations. Plants possess two mechanisms for the synthesis of methionine but the pathway shown in detail in Fig. 3–2 is believed to be the most important. Briefly, cysteine reacts first with phosphohomoserine (a high-energy amino acid analogous to O-acetylserine) to form cystathionine. Note that in this reaction cysteine is joined to homoserine via its sulphur atom; the hydrolysis of the phosphate group from phosphohomoserine ensures that the reaction is unidirectional. In the second reaction cystathionine is cleaved into two fragments by lysis of the —C—S— bond adjacent to that formed in the previous reaction; ammonia is also produced in the reaction. The net effect of the two reactions therefore is to transfer the sulphur atom of cysteine to homoserine to make homocysteine. Finally, a methyl group is added to homocysteine from a suitable methyl-group donor, thus completing the synthesis of methionine.

Our understanding of sulphate assimilation by plants explains certain metabolic changes associated with sulphur deficiency in plants. Deficient plants contain less cysteine, homocysteine and methionine relative to the other amino acids. They also contain less protein but more serine and homoserine. Any of these indices could be used to ascertain whether a plant is sulphur-deficient, but the most popular method is to determine whether the growth of plants on test plots is enhanced by the addition of sulphate. Another commonly used method is to measure the total amount of sulphur (regardless of form) in plant material. In general, plants containing less than 2 mg of sulphur per gram of dry leaf material will respond to applications of sulphate but the exact figure varies with the species, the age of the leaves analysed and several other variables.

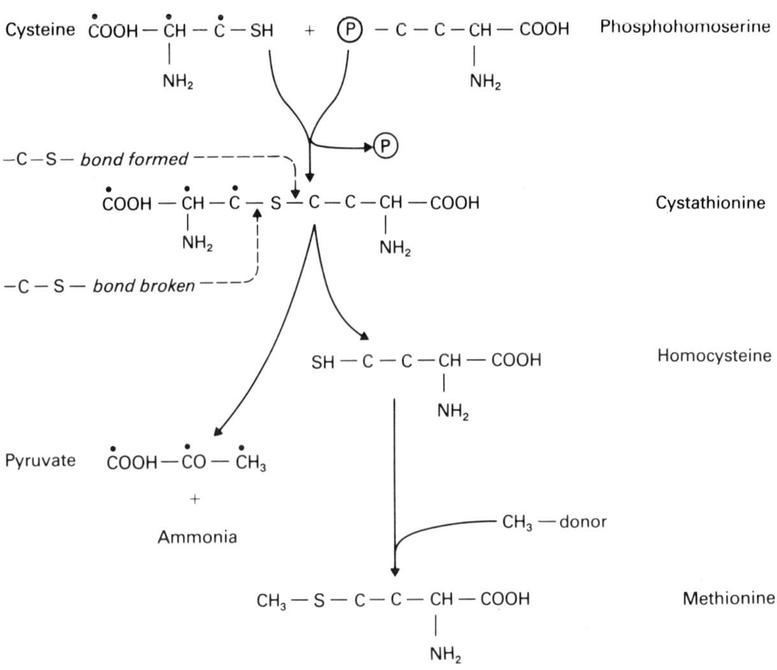

Fig. 3–2 Pathway for the synthesis of methionine from cysteine in plants and other organisms which catalyse assimilatory sulphate reduction. Only the carbon skeletons of the various compounds are shown. The symbol Ⓟ refers to a phosphate group and carbon atoms arising from cysteine are shown as $\dot{C}$.

3.3 Sulphur nutrition of monogastric animals

Monogastric animals possess a single stomach. Ingested food passes directly to this organ by way of the oesophagus. The stomach and the duodenum are the major sites of hydrolysis (digestion) of large polymeric compounds such as starch and protein to their constituent monomeric units. These are subsequently absorbed via the intestines. The hydrolytic processes in the stomach are achieved through the combined action of hydrolytic enzymes secreted into the stomach and the highly acid conditions maintained within it.

Monogastric animals do not possess mechanisms for the synthesis of cysteine and methionine from inorganic sulphur. Normally, the protein fraction of ingested foods is the most important source of these amino acids. The proteolytic enzymes of the stomach and duodenum (pepsin, trypsin and chymotrypsin) hydrolyse protein into its component amino acids. These are subsequently absorbed from the intestines into the bloodstream and transported to those sites in the animal where they are required for the synthesis of protein, coenzyme A, etc.

The sulphur metabolism of animals is in some respects the opposite of that in plants, though the mechanisms by which this is achieved are completely different – indeed they would have to be since, as noted in section 3.2, the synthesis of cysteine and methionine in plants are essentially irreversible processes. The important features of sulphur metabolism in animals are shown in Fig. 3–3. It is apparent that animals can synthesize cysteine from methionine (but not vice versa) and that they oxidize cysteine to inorganic sulphate. Accordingly, all of the animals requirements for sulphur can, in theory, be supplied by methionine alone, and for this reason, methionine (but not cysteine) is known as an essential amino acid. If methionine is the only source of sulphur supplied,

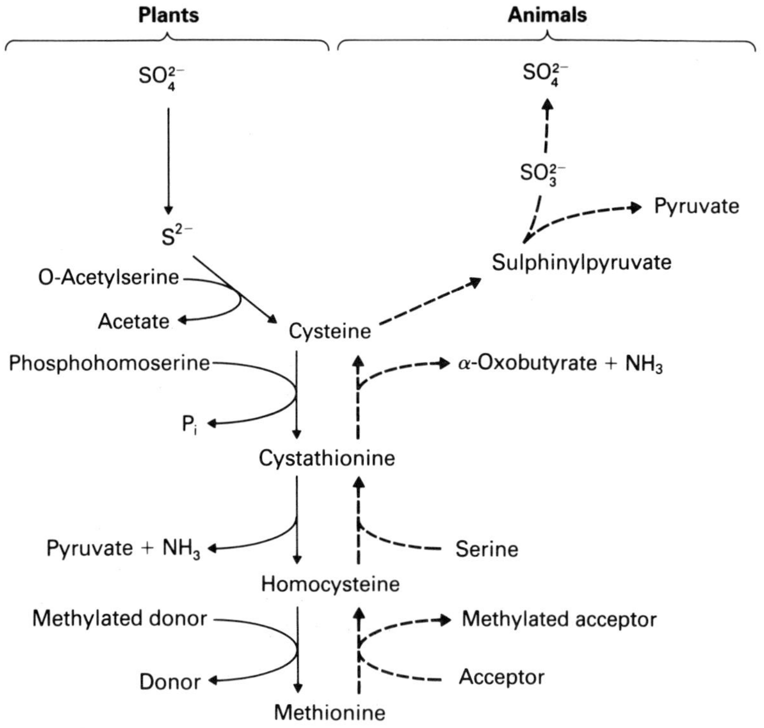

Fig. 3–3 Comparison of sulphate assimilation in plants with methionine metabolism in animals. Plant processes are shown on the left with continuous arrows, those in animals on the right with broken arrows. Note that the intermediates in the formation of methionine from cysteine in plants are common to those in the synthesis of cysteine from methionine in animals but that the mechanisms and bioenergetic principles involved are quite different.

some of it will be used for the synthesis of cysteine. It follows that if cysteine is also present in the diet the amount of methionine required will be proportionately less but clearly the methionine requirement of animals cannot be met by cysteine alone. In general, approximately 60–80% of an animal's requirement for methionine can be replaced by cysteine. Adult men require a combined intake of methionine and cysteine of approximately 1.1 g day^{-1}; women require somewhat less (0.55 mg day^{-1}).

3.4 Methionine and cysteine content of foodstuffs

The absolute requirement of monogastric animals for methionine raises the question of the methionine content of foodstuffs that they consume. However, it must also be borne in mind that animals also require eight other essential amino acids in addition to methionine and that cysteine can, in part, substitute for methionine. Animals require more of some of the essential amino acids than they do of others; relative amounts of the essential amino acids required by man are shown in Table 7. The table

Table 7 Provisional estimates of the amounts of the essential amino acids required in a protein to enable all the amino acids in the protein to be completely utilized in human nutrition. (From *Energy and Protein Requirements*, FAO Nutrition Meetings Report Series 52, 1973.)

Amino acid	*Amount* (mg g^{-1} *of protein nitrogen*)
Isoleucine	250
Leucine	440
Lysine	340
Methionine plus cysteine	220
Phenylalanine plus tyrosine	380
Threonine	250
Tryptophane	60
Valine	310

includes an entry 'methionine plus cysteine'. This is because 60–80% of the methionine requirement of animals can be replaced by cysteine. In most foodstuffs the amount of sulphur in cysteine is considerably less than 60% of the total amount of sulphur in cysteine plus methionine. Therefore cysteine can be included in the calculations of the requirement for the sulphur amino acids.

The best source of the essential amino acids would be a foodstuff rich in protein containing the essential amino acids in the proportions shown in Table 7. Human milk and whole eggs have these properties. Foodstuffs with a relative deficiency of an essential amino acid have a low biological or nutritional value since a certain proportion of the other amino acids, which might be present in desirable proportions, are not fully utilized by the animal.

The other important aspect in considering the methionine and cysteine content of a foodstuff is the amount of protein it contains; a foodstuff containing a desirable balance of the essential amino acids is hardly satisfactory if it contains very small quantities of protein (e.g. apples and pears). Typical protein, sulphur amino acid and lysine contents of some raw foodstuffs consumed by man and his domestic animals are shown in Table 8. It is evident that fruit and vegetables contain very low concentrations of protein and consequently very little methionine and cysteine. To satisfy his needs for sulphur an adult man would need to consume approximately 35 kg of apples per day, surely a daunting prospect even for the most avid apple-lover! Fruit and vegetables cannot

Table 8 Typical protein, methionine, cysteine and lysine contents of some common unprocessed foodstuffs. Most fruits (values not shown in the table) contain less than 10 g protein kg^{-1} fresh weight. The relevant FAO values required for efficient utilization of all the essential amino acids are shown at the bottom of the table. (Values abstracted from ALTMAN, P. L. and DITTMER, D. S. (eds.), (1968). *Metabolism*, Federation of American Societies for Experimental Biology, Bethesda, Maryland.)

Foodstuff	*Protein* ($g\ kg^{-1}$ *fresh weight*)	*Amino acid composition of protein* ($mg\ g^{-1}$ N) *Methionine*	*Cysteine*	*Methionine plus cysteine*	*Lysine*
Vegetables					
cabbage	13	58	125	183	295
potatoes	21	78	60	138	333
Cereals					
rice	75	107	81	188	235
wheat	135	89	128	217	160
Legumes					
french beans	223	63	62	125	464
mung beans	242	68	39	107	427
Meats					
beef	169	155	79	234	546
chicken	234	163	84	247	549
Other animal products					
cow's milk	35	156	57	213	496
whole eggs	129	196	146	342	400
Provisional FAO values (see Table 7)				220	340

therefore be viewed as important sources of sulphur for the human diet although they do have other important nutritional qualities. Even potatoes, which have in the past formed the staple diet of the inhabitants of some countries, are poor sources of sulphur amino acids; they contain little protein, being mostly comprised of starch, and the proportion of the sulphur amino acids in the protein is too low for efficient utilization of the other essential amino acids.

The most satisfactory sources of sulphur amino acids for human consumption are meat and the animal products milk, cheese and eggs. These commodities contain high concentrations of protein and the essential amino acids are present in approximately the right proportions for maximum utilization of the protein. Man's needs for sulphur can therefore be readily satisfied by a small daily intake of foodstuffs of animal origin. However, whilst this would appear to be the simplest means of fulfilling man's daily need for sulphur, this is not always possible for a variety of economic, cultural, ethnic and geographic reasons. One of the most important considerations involved here is that the conversion of plant foodstuffs to animal foodstuffs is only about 10% efficient. Clearly, if food is scarce, a given amount of a plant foodstuff will feed far more people than if it is converted to animal products. It follows that plant foodstuffs are also much cheaper. For these reasons more than any other, the inhabitants of many nations use the seeds of cereals and legumes as direct sources of energy (starch and fat) and protein. Cereal seeds, however, are not particularly rich sources of protein and they contain inadequate concentrations of lysine and, to a lesser extent, the sulphur amino acids (compare Tables 7 and 8). For this reason it is not uncommon to supplement cereal foodstuffs for pigs and poultry raised under intensive conditions with fishmeal or meatmeal which are cheap sources of protein rich in lysine. Supplements of this kind, however, are not particularly attractive to man. One approach to this problem is to seek a plant foodstuff with a high protein content containing adequate amounts of lysine. Legume seeds have these properties but they contain inadequate amounts of the sulphur amino acids (Table 8). By careful supplementation of a cereal foodstuff with leguminous seed it is possible to devise a diet which approximately fulfils the requirements of mono-gastric animals for both energy and lysine.

3.5 Increasing the world supply of sulphur amino acids

The supplementation of cereal foodstuffs with legume seed corrects the lysine deficiency of cereal protein, but because cereal seed is marginally deficient in sulphur amino acids and legumes highly deficient, this leaves a deficiency of sulphur amino acids in a mixture of the two.

One way to approach this problem is to study the individual proteins that collectively comprise the protein complement of legume seeds.

Leguminous seeds contain a large number of proteins (including the enzymes and structural proteins characteristic of any organism) but the bulk of the protein is contained in a few storage proteins which are utilized for energy and for the synthesis of new proteins when the seed germinates. Storage proteins, like all proteins, have a fixed amino acid sequence and therefore a fixed amino acid composition. It follows that if a particular amino acid is in short supply during the synthesis of a protein, then synthesis of the protein in question will cease. One legume seed in which the various storage proteins have been examined in relation to sulphate nutrition of the parent plant is *Lupinus angustifolius* (blue lupin). Seeds of this species contain the storage proteins conglutins α, β and γ. Conglutin γ is relatively rich in the sulphur amino acids compared to conglutin β (Table 9). Conglutin α is intermediate between the two.

Table 9 Relative amounts of some amino acids in the storage proteins conglutins α, β and γ from the seeds of *Lupinus angustifolius*. Conglutin γ is a sulphur-rich protein, conglutin β is sulphur-poor and conglutin α is intermediate. The results are expressed as moles of amino acid per 100 moles of total amino acids for each protein. (From BLAGROVE, R. J. and GILLESPIE, J. M. (1975). *Aust. J. Plant Physiol.*, **2**, 13–27.) Approximate estimates of the relevant FAO values (see Table 7) expressed in the same unit are also shown.

	Moles of amino acid per 100 moles total amino acids			
Amino acid	*Conglutin* α	*Conglutin* β	*Conglutin* γ	*FAO value*
Cysteine	0.66	0.30	2.91	2.3 (Cysteine + Methionine)
Methionine	0.17	0.00	1.02	
Lysine	2.82	3.88	4.10	3.2
Histidine	1.87	1.52	4.58	
Arginine	9.04	9.48	2.80	
Glutamate	23.60	20.80	8.33	

When plants are grown at various levels of inorganic sulphate it is found that conglutins α and γ are virtually absent in the seeds of plants grown at low concentrations of sulphate but conglutin β is readily detected (Fig. 3–4). Although the level of sulphur nutrition cannot affect the proportion of the amino acids in particular proteins it is apparent that the amount of the sulphur-rich proteins synthesized by *Lupinus* is highly dependent on sulphur nutrition which affects the sulphur amino acid content of the seed and therefore its nutritional quality.

The study of legume proteins has proved important in another sense too; the development of legume varieties with the genetic capacity to synthesize relatively more of the sulphur-rich storage proteins has become an important goal for the plant geneticist. In fact the

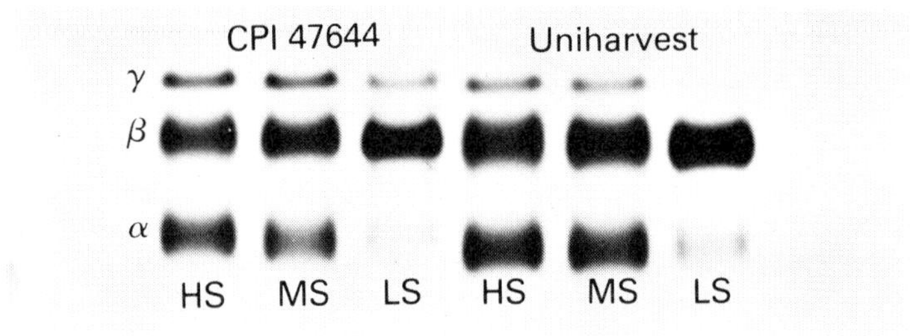

Fig. 3–4 Electrophoretogram showing the relative amounts of the storage proteins conglutins α, β and γ in the seeds of two varieties of *Lupinus angustifolius* when the parent plants were grown at high (HS), medium (MS) and low (LS) concentrations of inorganic sulphate (2.25 mM, 0.375 mM and <0.06 mM respectively). (From BLAGROVE, R. J., GILLESPIE, J. M. and RANDALL, P. S. (1976). *Aust. J. Plant Physiol.*, 3, 173–84.)

composition of the essential amino acids in conglutin γ from *L. angustifolius* is equal or even superior to the composition recommended by FAO (Tables 7 and 9).

3.6 Sulphur nutrition of ruminant animals

Ruminant animals (e.g. sheep, cow, goat) are characterized by a forestomach comprised of three organs: rumen, reticulum and omasum. These organs are interposed between the oesophagus and true stomach (referred to as abomasum in ruminants). The rumen is by far the largest organ of the forestomach, virtually filling the left side of the abdominal cavity. The rumen, unlike the stomach of monogastric animals and the abomasum of ruminants, is not maintained at a highly acid pH. Further, hydrolytic enzymes, other than those which enter with saliva, are not secreted into it. As such, it is a perfect haven for the growth of various species of microorganisms and, in the nutritional sense, it can be viewed as an anaerobic fermentation tank for the digestion of plant material into forms which can be absorbed by the host. Microorganisms in the rumen (bacteria and protozoa) ferment cellulose and starch to short-chain fatty acids such as acetate, propionate and butyrate. These are absorbed through the wall of the rumen and transported to the liver where they are used for the synthesis of sugars and fats. Thereafter the carbon metabolism of ruminants is essentially the same as for monogastric animals.

Given the large number of different species of microorganisms in the rumen, what then is the fate of the sulphur amino acids present in ingested plant material? If an aliquot of rumen contents (containing microorganisms) is removed from the rumen and incubated with cysteine it is found that the carbon atoms of cysteine are recovered in the short-chain fatty acids described above but the sulphur atom is released as inorganic sulphide. The same process occurs *in vivo* in the rumen. Methionine is metabolized in an analogous manner except that the process is somewhat slower. The ruminant animal cannot itself synthesize the sulphur amino acids from short-chain fatty acids and sulphide and the microorganisms of the rumen have destroyed the very compounds required for ruminant nutrition! Obviously, some mechanism must exist for the resynthesis of cysteine and methionine. The processes involved here can be understood from experiments on the growth of sheep and rumen microorganisms when supplied with a purified diet containing negligible amounts of organic sulphur with or without added sulphate. Sheep fed on a diet without added sulphate fail to grow and some deaths occur, but animals receiving sulphate grow normally. Similarly, if samples of rumen contents from normal sheep are incubated with the purified foodstuffs, microbial growth is far greater in the presence of inorganic sulphate. The latter experiment implies that the rumen microorganisms are able to use inorganic sulphate for growth, i.e. they synthesize the sulphur amino acids and incorporate them into microbial protein. Anaerobic dissimilatory sulphate-reducing bacteria (e.g. *Desulfotomaculum ruminis*) are invariably present in rumen contents. They produce sulphide from sulphate and this in turn is used by other microorganisms for the synthesis of sulphur amino acids. The latter group of organisms in return supply suitable carbon substrates which are used as an energy source by the sulphate-reducing organisms during sulphide production. Cysteine and methionine are metabolized in the rumen somewhat differently; these compounds are broken down to sulphide and short-chain fatty acids. The sulphide is then re-incorporated into sulphur amino acids by other microorganisms. In other words, the sulphur amino acids are only free from microbial degradation in the rumen once they are synthesized within a microbial cell and incorporated into microbial protein.

How the rumen microflora can synthesize microbial protein from inorganic sulphur, cysteine, and methionine has been discussed but there remains the matter of the sulphur nutrition of the ruminant itself – a relatively simple matter. Microbial cells are continually drained out of the rumen via the other organs of the forestomach to the abomasum, the physiology of which can be likened to the stomach of monogastric animals. Microbial cells are subjected to the processes of acid digestion and proteolytic hydrolysis by enzymes secreted into the abomasum. The sulphur amino acids which are released by these processes are subsequently absorbed from the intestines. The passage of sulphur in the alimentary system of a ruminant animal is shown diagrammatically in Fig. 3–5.

The sulphur nutrition of the ruminant animal provides an elegant

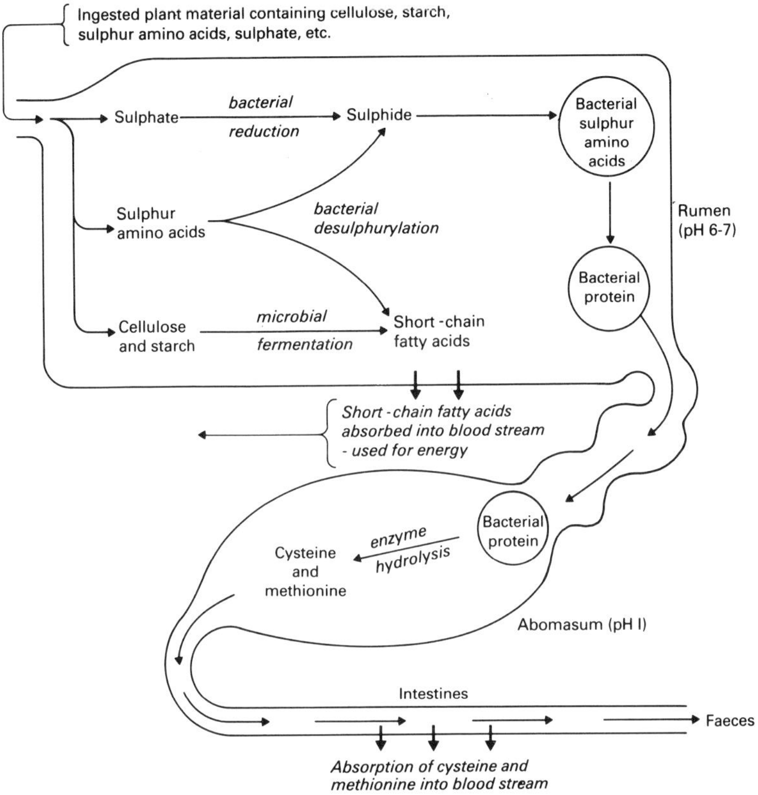

Fig. 3–5 Sulphur nutrition of a typical ruminant animal.

demonstration of symbiosis at two distinct levels. Firstly, the rumen microorganisms are dependent on each other for the supply of sulphide for amino acid synthesis and for short-chain fatty acids for sulphide production. Secondly, the microbial population as a whole is dependent on the host for a supply of water and plant material and an anaerobic environment for microbial growth which is ultimately used by the animal for its own nutrition.

The ability of ruminant animals to use inorganic sulphur as a sulphur source is probably of practical as well as academic interest. If the herbage ingested by ruminants is low in total sulphur (sulphur amino acids and inorganic sulphur) this can, in theory, be simply rectified by supplying the animal with inorganic sulphur. This has been accomplished in laboratory experiments and might in due course prove to be commercially important.

4 Cycling of Sulphur

With few exceptions, organisms are localized on the terrestrial surface of the earth, in the soil which forms the outer few metres of the earth's crust, and in the water of lakes, rivers and oceans. Some organisms also spend some portion of their life cycle in the atmosphere. These regions are collectively known as the biosphere. Whilst sulphur occurs in high concentrations in localized deposits (e.g. FeS, S^0, CuS) in various parts of the world, these sources are not usually located within the biosphere (e.g. the elemental sulphur deposits of Texas, U.S.A.). Gypsum ($CaSO_4$) is an important exception but deposits of this mineral usually occur in arid and/or saline environments which are not particularly favourable for the growth of organisms. In general the amount of sulphur in the terrestrial component of the biosphere in relation to the annual amount consumed by terrestrial organisms is very small. For example, a typical soil used for growing a pasture to support sheep for wool production contains 400 kg sulphur per hectare in the top 25 cm. On the other hand the pasture plants typically remove 40 kg of sulphur per year from the soil. Continuous removal of sulphur at this rate would exhaust the sulphur in the top 25 cm of soil in 10 years! Cycling of sulphur is therefore essential for continued biological activity on the earth.

Some organisms which catalyse various interconversions of inorganic and organic sulphur have been described in earlier chapters. Collectively, the processes catalysed by these organisms largely account for the cycling of sulphur within the biosphere. Global cycling of sulphur on the other hand is more concerned with the cycling of sulphur between the biosphere and the sulphur reserves. Typical of the processes involved here are the removal of sulphur from the biosphere in sediments and additions from volcanoes and the weathering of rocks. Redistribution of sulphur over the earth by way of wind and waterflow are also important component processes of global cycling. But the cycling of sulphur also occurs on a less extensive scale. Indeed, most ecosystems have their own characteristic sulphur cycle (often involving only a few of the groups of organisms described in earlier chapters) which maintains a steady state level of sulphur in the ecosystems. The continued removal of biological material from agricultural ecosystems by harvesting, however, entails loss of sulphur from the ecosystem. This practice will, sooner or later, lead to decreased productivity unless steps are taken to replace the sulphur removed.

4.1 The sulphur reserves of the world

The sulphur reserves of the world are not generally available for short-term cycling by organisms. An assessment of these reserves, however, is important for two reasons. First, as mentioned in the introduction to this chapter, they are important components of the global sulphur cycle. Secondly, twentieth-century man is continually removing prodigious amounts of sulphur from these reserves and releasing it into the biosphere. The world's sulphur reserves are obviously very difficult to assess but an estimate of these and the form in which they occur is shown in Table 10. Contrary perhaps to popular belief, coal, oil shales and

Table 10 Estimates of the world reserves of sulphur. (Reproduced from BIXBY, D. W., *The Role of Phosphorus in Agriculture*, 1978, by permission of the American Society of Agronomy and Soil Science Society of America, Madison.)

	Sulphur (10^9 tonnes)	
Reserve	*Identified and recoverable*	*Probable*
Elemental sulphur	0.53	?
H_2S in sour natural gas	0.15	0.9
Petroleum	0.02	1.59
Iron pyrites (FeS)	0.56	?
Volcanic rocks	0.10	0.13
Metal sulphides other than FeS	0.28	?
Gypsum ($CaSO_4$)	?	(substantial)
Tar sands	?	> 1.8
Coal	?	221
Oil shale	?	281
Sea water (as sulphate)	?	(vast)

gypsum ($CaSO_4$) are quantitatively the most important reserves of sulphur. Elemental sulphur (S^0) and iron pyrites (FeS) are the forms most actively exploited by man for industrial purposes; they form only a small part of the total sulphur reserves.

4.2 Anthropogenic additions of sulphur to the biosphere

Man consumes the sulphur reserves in two ways. The first involves the deliberate extraction of elemental sulphur and pyrites for the manufacture of various sulphur-containing commodities required by man. Most of the sulphur extracted from these sources is used for the manufacture of fertilizers (e.g. superphosphate, and ammonium sulphate) and is therefore fed directly into the biosphere. This amounts to about 30 million tonnes annually and is referred to as voluntary sulphur production. But man also consumes the sulphur reserves through his consumption of coal, natural gas and petroleum and to a lesser extent by the refining of base metal sulphide ores. Some estimates of the consumption of sulphur from these reserves are shown in Table 11. Most of

Table 11 Some estimates of man's consumption and recovery of sulphur from the world's reserves. (Values abstracted and recalculated from ROBINSON, E. and ROBBINS, R. C. (1968). *Stanford Research Institute* Project No. PR-6755, and BIXBY, D. W., *The Role of Phosphorus in Agriculture*, 1978, by permission of the American Society of Agronomy and Soil Society of America, Madison.)

	Sulphur (10^6 tonnes year^{-1})	
Source	*Consumed*	*Recovered*
Elemental sulphur	18.0	18.0
Iron pyrites	11.0	11.0
Petroleum		
(a) refining	16.9	
(b) combustion	10.3	
Coal	46.0	
Smelting		22.0 (Petroleum to Natural gas, bracketed)
(a) copper	5.8	
(b) lead	0.7	
(c) zinc	0.6	
Natural gas	>6.7	
Total	106.0	42.0

the sulphur from these sources is released into the atmosphere as gaseous SO_2, the major contributions coming from the oxidative combustion of coal and petroleum. Anthropogenic emissions of SO_2 rapidly enter the biosphere; SO_2 is either directly absorbed by plants, or oxidized to SO_3 in the atmosphere and returned to the soil in rainwater as inorganic sulphate. As discussed in section 5.1, SO_2 is a serious pollutant. The important point to note here, however, is that anthropogenic emissions of SO_2 are very substantial and although they are transitorily toxic they nevertheless represent a significant addition of sulphur to the biosphere.

In recent times environmental protection agencies have introduced legislation to lessen the discharge of SO_2 into the atmosphere. This in turn has led to the removal of sulphur from gaseous industrial wastes or from materials which would in due course give rise to them. Typical of these processes is the Claus sulphur-recovery plant for the removal of H_2S from sour natural gas. H_2S is trapped in an alkali (usually ethanolamine) and then released by heating. Sufficient oxygen is then added so that one third of the H_2S is oxidized to SO_2 when the gas mixture is combusted. SO_2 is then reacted with the remaining two thirds of H_2S at high temperatures in a catalytic converter to produce elemental sulphur:

$$H_2S + 1\tfrac{1}{2}O_2 \longrightarrow SO_2 + H_2O$$
$$2H_2S + SO_2 \longrightarrow 3S^0 + 2H_2O$$
$$\text{Sum: } 3H_2S + 1\tfrac{1}{2}O_2 \longrightarrow 3S^0 + 3H_2O$$

Sulphur is also recovered from other gaseous sulphur emissions by a variety of processes. Collectively, the sulphur produced in this way is referred to as involuntary sulphur production. At present, less than 30% of the total potential production of involuntary sulphur is realized; this proportion is likely to increase rapidly over the next ten to fifteen years. In this event the combined output of voluntary and involuntary sulphur production is likely to exceed man's requirements for sulphur. It remains to be seen whether future economic, environmental, social and industrial factors will force man to stockpile his production, consume it in new processes (e.g. road making), cease voluntary sulphur production or return it to the biosphere in fertilizers.

4.3 Additions of sulphur to the biosphere by natural physical processes

Sulphur is emitted in various forms from volcanoes, geysers and springs. Presumably most of this sulphur enters the biosphere. Probably more important, however, is the constant release of sulphur into the soil by the weathering of rocks containing sulphur minerals such as pyrites (FeS), galena (PbS), gypsum ($CaSO_4$), etc. The rate of weathering is dependent on many factors but it is probable that the rate at, or close to, the earth's surface is accelerated by various biological processes such as the secretion of acids by organisms (e.g. lichens) and the opening of rock fissures by the roots of plants, especially trees. The rate of addition of sulphur to soil by weathering has been estimated to be about 10% of that of sulphur accessions from the atmosphere.

4.4 Removal of sulphur from the biosphere

Removal of sulphur from the biosphere is accompanied by a gain in the sulphur reserves. Most of the processes which contribute to the removal of sulphur involve combination of biological and physical factors. Coal, natural gas and petroleum, which represent a large proportion of the sulphur reserves, are formed from biological material. Presumably the sulphur associated with these sources is due to the sulphur content of the biological material from which they are made. The formation of the fossil fuels, however, also required suitable physical conditions of temperature, pressure, time and anaerobic conditions; the latter condition is responsible for the retention of carbon and sulphur in these deposits in reduced forms. There is also some evidence that the deposits of elemental sulphur such as are found in the southern U.S.A., Sicily and the Carpathians are also due to the combined action of biological and physical factors. The biological processes involve reduction of sulphate to sulphide by dissimilatory sulphate reduction and partial oxidation of sulphide to elemental sulphur by photosynthetic sulphur-oxidizing bacteria. Both component processes require anaerobic conditions:

Dissimilatory bacteria $SO_4^{2-} + 2[CH_2O] \longrightarrow S^{2-} + 2CO_2 + 2H_2O$
Photosynthetic bacteria $2H_2S + CO_2 \xrightarrow{h\nu} [CH_2O] + 2S^0 + H_2O$

Once again suitable geological conditions are required to isolate the sulphur produced in these reactions from the biosphere.

A third example of sulphur removal from the biosphere is the formation of insoluble metal sulphides. The processes of dissimilatory sulphate reduction and desulphurylation lead to the production of sulphide. This reacts with the cations of base metals to form the corresponding sulphides which, because of their low solubility products, precipitate out of solution. There is growing evidence that at least some deposits of metal sulphide ores (especially iron pyrites, FeS) are of a sedimentary nature. Further, it is well known that the beds of lakes and oceans are anaerobic and favourable for the formation of sulphide by biological processes. Trace amounts of base metal cations (e.g. Fe^{2+}, Zn^{2+}, Cu^{2+}) in the oceans would therefore cause removal of sulphur from the biosphere and perhaps initiate the formation of sedimentary metal sulphide deposits.

4·5 Volatilization of sulphur by organisms

Both the desulphurylating and dissimilatory sulphate-reducing bacteria produce H_2S, some of which, depending on the nature of the habitat, escapes into the atmosphere. In the case of shallow anaerobic muds and marshes receiving high inputs of sulphate, losses by volatilization can be very high. Higher plants also volatilize sulphur. The latter process can be demonstrated as follows. Plants are enclosed in a chamber and grown in a nutrient solution containing sulphate labelled with the radioactive isotope ^{35}S. Air is passed into the chamber and, after circulating around the plants, is passed over a series of absorbent materials (e.g. charcoal) before allowing it to escape. When the absorbent materials are examined at the completion of the experiment they are found to contain ^{35}S-label. Since negligible quantities of label are recovered in the traps when plants are not placed in the chamber, this implies that plants volatilize sulphur. However, the biological significance of this process in plants is not understood. Similarly very little is known about the form in which sulphur is volatilized although at least some of it is lost as H_2S. Some plants, especially those belonging to the family Cruciferae, volatilize large amounts of sulphur; anyone who has passed by a field of cabbages just prior to harvest could not fail to detect it! Nevertheless, the sulphur volatilized by organisms does not really represent loss of sulphur from the earth's surface as it is returned to the soil in rainfall and by direct absorption.

4.6 Global cycling of sulphur

Global cycling of sulphur occurs at two levels. The first of these concerns the addition of sulphur from the sulphur reserves to the biosphere and vice versa. These processes, described in sections 4.2 to 4.4 are shown by the dotted lines in Fig. 4–1. The second level of global sulphur cycling concerns the interconversions of sulphur on the earth's surface (land and water) which in turn are in equilibrium with the sulphur in the atmosphere. These processes are shown by the continuous lines in Fig. 4–1. The interconversions on the earth's surface are essentially biological. Some sulphur is lost through the activity of organisms to the atmosphere, but this is recycled back to the earth's surface. This emphasizes the importance of the atmosphere in the cycling of sulphur. The atmosphere also plays an important part in redistributing sulphur over the earth's surface.

It should be emphasized that the component steps of the sulphur cycle shown in Fig. 4–1 apply to the world as a whole. Some of the component processes can be very localized. There are, for example, relatively few sites in the world conducive to large-scale oxidation of H_2S by anaerobic photosynthetic bacteria. Similarly, the anthropogenic emissions of SO_2 into the atmosphere come from point sources concentrated in the major industrial centres of the world. Conversely, organisms catalysing assimilatory sulphate reduction, desulphurylation and chemoautotrophic oxidation of reduced sulphur are active in most regions of the biosphere (soil and water) and these processes account for most of the biological cycling of sulphur. Various factors can cause the redistribution of sulphur over the globe; natural drainage (especially surface run-off), water, wind and various weather conditions can serve both to lower the concentration of sulphur at sites of high concentration and increase the sulphur content of regions with a low concentration. In this respect, the addition of sulphate to the atmosphere from sea-spray is particularly important in

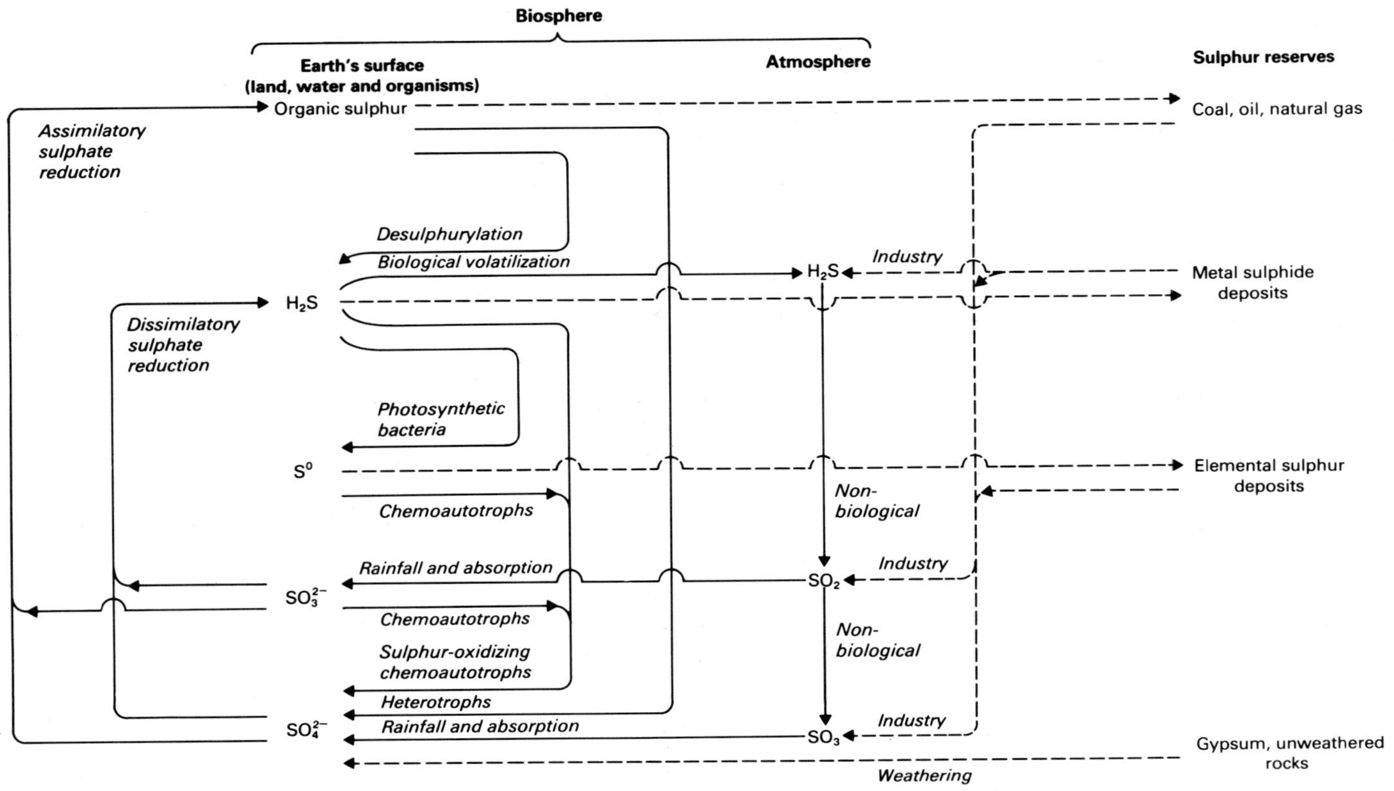

Fig. 4–1 The global sulphur cycle. The continuous lines represent short-term sulphur cycling within the biosphere and atmosphere. The dotted lines show long-term cycling between the components of the biosphere and the sulphur reserves.

transfering sulphur from oceans to the land. Some estimates of the rates of some component processes of the global sulphur cycle are shown in Fig. 4–2.

One important matter which is relevant here is the effect of

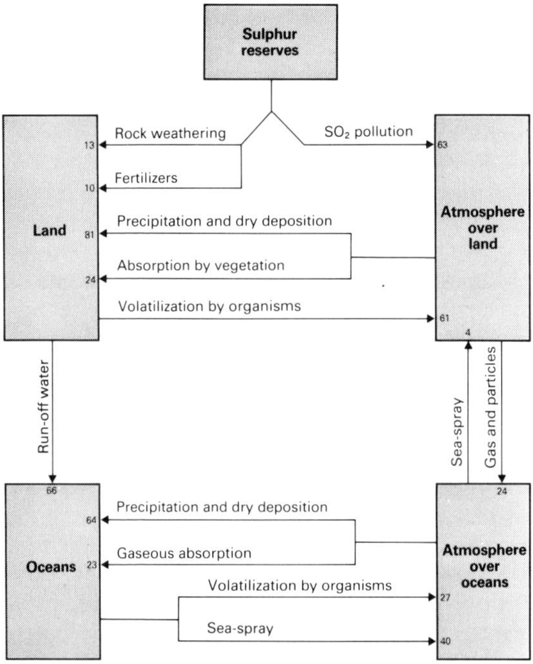

Fig. 4–2 Estimates of sulphur flux between the major components of the global sulphur cycle. Values are expressed in millions of tonnes of sulphur per year. (Redrawn and recalculated from ROBINSON, E. and ROBBINS, R. C. (1968). *Stanford Res. Inst.* Report no. PR-6755.)

anthropogenic additions of sulphur on the biosphere and the global sulphur cycle. In the short term, these are likely to accumulate in the soil and increase primary productivity, a matter which will be discussed in section 5.3. It can be predicted from Fig. 4–2 that in the long term the rate of removal of sulphur from the biosphere will increase and that it will accumulate on the floors of lakes and oceans as organic matter and sulphides and so return to the sulphur reserves. Only subsequent, long-term geological events could make the sulphur accessible for re-exploitation.

4·7 Sulphur cycling in aquatic systems

Lakes and ponds provide excellent examples of sulphur cycling on a localized scale. Sulphur from the surrounding land enters the pond in natural drainage water and wind-blown material. These materials contain both inorganic sulphur (mostly sulphate) and organic sulphur associated with dead biological material (organic matter). These additions are greatest in the autumn (e.g. leaf litter from plants). In small ponds the input of organic matter causes the growth of heterotrophic bacteria which rapidly consume the oxygen in the water causing anaerobic conditions in all but the top few millimetres of water. The anaerobic conditions, together with the input of sulphate, provide ideal conditions for the growth of dissimilatory sulphate-reducing bacteria thereby causing the sulphide concentration to increase. Desulphurylation of protein present in the organic matter also contributes to sulphide production. Water at the surface, however, contains some oxygen and is therefore suited to the growth of sulphide-oxidizing chemoautotrophic bacteria: a scum of these organisms together with the sulphur granules (S^0) that they produce can frequently be seen on the surface. The formation of elemental sulphur is also aided by photosynthetic sulphide-oxidizing bacteria which grow below the surface of the pond where it is anaerobic but sufficiently shallow for sunlight to penetrate the water. Thus the pond can establish a biological sulphur cycle involving only bacteria, i.e. oxidation by chemoautotrophs and photoautotrophs at (or near) the surface, and production of sulphide by organisms catalysing dissimilatory sulphate reduction and desulphurylation lower in the water profile. The autotrophs near the surface of the pond assimilate CO_2 and so form organic compounds which, on the death of these organisms becomes available as an energy source for dissimilatory sulphate-reducing and desulphurylating bacteria (Fig. 4–3). Thus the bacterial system is self-sustaining provided it receives inputs of oxygen, CO_2 and light for the two types of autotrophs and that the other elements required for bacterial growth (N, P, K, etc.) are also available. Such a system is referred to as a sulphuretum.

Sulphur cycling in the Black Sea is a slightly more complex variation of the simple bacterial sulphuretum. In this case the bacterial inter-conversions of sulphur are displaced downwards in the water profile because the aerobic layer at the surface is much deeper due to vigorous aeration by wave turbulence. The greater depth of aerated water provides conditions suitable for the growth of planktonic algae which in turn produce even higher oxygen concentrations. The organic matter produced by the algae is oxidized by a mixed population of aerobic heterotrophs which grow immediately below the algae and deplete the water of oxygen in the layer in which they grow. Below this again the processes are essentially as for the layered structures of the simple

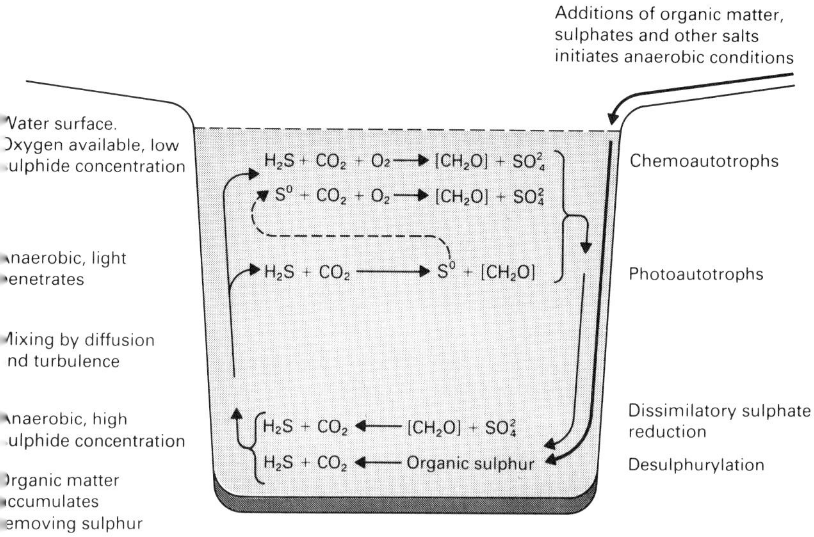

Fig. 4–3 Cycling of sulphur in a sulphuretum.

sulphuretum except that the photosynthetic sulphide-oxidizing bacteria are usually absent as the anaerobic zone is so far below the surface that the light intensity is not sufficient for their growth. Chemoautotrophic sulphide-oxidizing bacteria are found in the profile where sulphide diffusing upwards mixes with oxygenated water diffusing downwards.

4.8 Sulphur cycling in pastures and livestock

In some parts of the world that are used for primary production, plant and animal growth is limited by the availability of sulphur in the soil. Productivity in these areas can be increased by applying fertilizers containing sulphur. It is therefore of both practical and biological interest to know as much as possible about sulphur cycling under these conditions. Some of the questions that are pertinent here are as follows. How much of the sulphur added as fertilizer is available to plants and the animals that graze on them? How much sulphur is consumed by the grazing animal and is returned to the pasture? How much of the sulphur recycles?

Much can be learnt about sulphur cycling in the soil–plant–animal system by the use of the radioactively labelled isotope ^{35}S. If fertilizer containing ^{35}S (e.g. $Ca^{35}SO_4$) is supplied to a pasture, the plant material subsequently becomes labelled. Stock grazing the labelled pasture retain only 10% of the label they ingest; the remaining 90% is excreted in the urine (typically 55%, mostly as sulphate) and the faeces (typically 35%, mostly as organic sulphur). In sheep, some of the ingested sulphur is incorporated into wool which contains large amounts of keratin, a protein with an unusually high proportion of cysteine units. It is therefore a simple matter to monitor sulphur incorporation in sheep by analysing samples of wool for ^{35}S-label at various times.

The overall process of sulphur cycling can be observed by modifying the experimental design. A small proportion of a pasture is treated with $Ca^{35}SO_4$ and sheep are allowed to graze randomly over the entire pasture. After about nine months the area to which the ^{35}S-label was originally supplied is fenced off and the sheep replaced by new animals which have never previously been exposed to ^{35}S-label. It is found that the fleeces of the new animals became labelled in the ensuing period. This experiment demonstrates that sulphur has been completely recycled by the soil–plant–animal complex. The ^{35}S-label has been taken up by plants from the area to which it was orginally applied, ingested by sheep and transported by them to other parts of the pasture where most of it is excreted in urine and faeces. The excreted sulphur then becomes available to plants at the new sites (probably by way of microbial transformations in the soil) and is ingested by the new set of animals and some of it is incorporated into wool.

Much more detail about the quantitative aspects of sulphur cycling can be obtained from experiments similar to those described above if samples of soil and pasture as well as wool, faeces and urine from the animal are collected at various times and analysed for both total (unlabelled) sulphur content and ^{35}S-label. The ratio of ^{35}S-label to the amount of unlabelled sulphur is referred to as the specific radioactivity. When ^{35}S-label is applied to a sample area it mixes with any unlabelled sulphur already present in the soil. Plants on the site at the time of application of the ^{35}S-label contain only unlabelled sulphur, but any sulphur they subseqently take up will be a mixture of the two forms. Making certain assumptions one would predict that the specific radioactivity of the plants on the treated area will gradually increase until it is approximately the same as the soil. Similarly, stock grazing on the treated area would also show a gradual increase in the specific activity of their wool although the maximum specific activity will be much less than that of the plants on the treated area because the animals are free to graze on pasture growing on the untreated portions of the paddock, thereby obtaining additional sources of unlabelled sulphur. Weighed against this, however, are several factors which will, in the longer term, cause a decrease in specific radioactivity of the wool. The most important of these is that sheep transport labelled sulphur from the treated sites and excrete it on untreated sites. The specific radioactivity of the sulphur excreted by the animals will in turn be decreased as it becomes mixed with unlabelled sulphur in the soil on the untreated site. Since the animals graze randomly over the paddock it is evident that, given more time, the specific radioactivity of the wool of animals and of plants on the treated site will decrease whereas the plants on the untreated sites will increase. Eventually, however, all the components in the system will come to a constant specific radioactivity indicating that the ^{35}S-label, originally applied to only a small area of the paddock, is in equilibrium with sulphur in plant material, soil and wool. The experimental data in Fig. 4–4 show that when $Ca^{35}SO_4$ is applied to

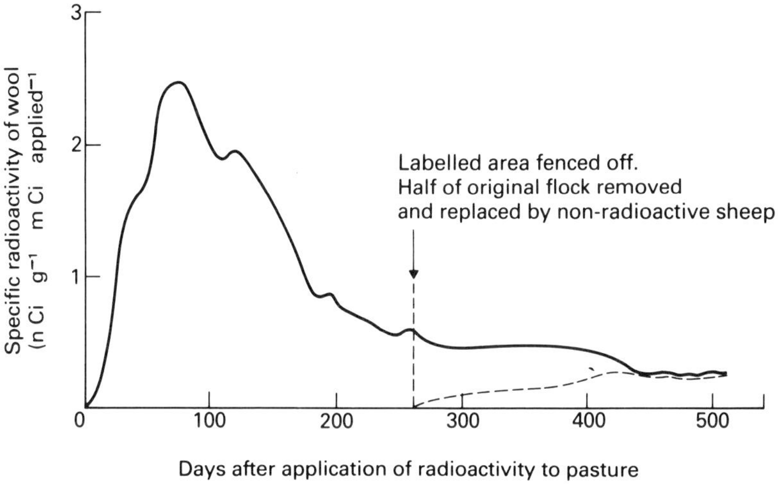

Fig. 4–4 Specific radioactivity of the wool of sheep (——) following the application of $Ca^{35}SO_4$ to approximately 0.015% of the area of the paddock in which they grazed. After 254 days the area to which the $Ca^{35}SO_4$ was originally applied was fenced off and half of the first group of animals removed and replaced with a second group (– – –) of non-radioactive animals. (From TILL, A. R., MAY, P. F. and McDONALD, I. W. (1970). In: MUTH and OLDFIELD (1970), 182–95.)

0.015% of the area of a pasture the specific radioactivity of the wool of sheep initially rises and then declines and approaches a constant specific radioactivity. When the treated area was fenced off after 254 days, the wool of newly introduced animals gradually approached the same specific radioactivity. Provided certain assumptions are made, these data can be used to calculate the amount of sulphur cycled from soil, via plants to the wool of sheep from the following relationship:

$$\text{Amount of S cycled from soil to wool} = \frac{\text{Amount of S applied to treated area} \times \text{Specific radioactivity of S applied}}{\text{Specific radioactivity of wool at equilibrium}}$$

Similar approaches can be used to calculate many parameters of sulphur cycling in the soil–plant–animal system including the rate of the partial processes. Typical estimates of some of these parameters are shown in Fig. 4–5. They show that about 60% of the total amount of sulphur in the complete system undergoes short-term cycling; the remaining 40% is locked away in forms which are not available to plants. At any one time,

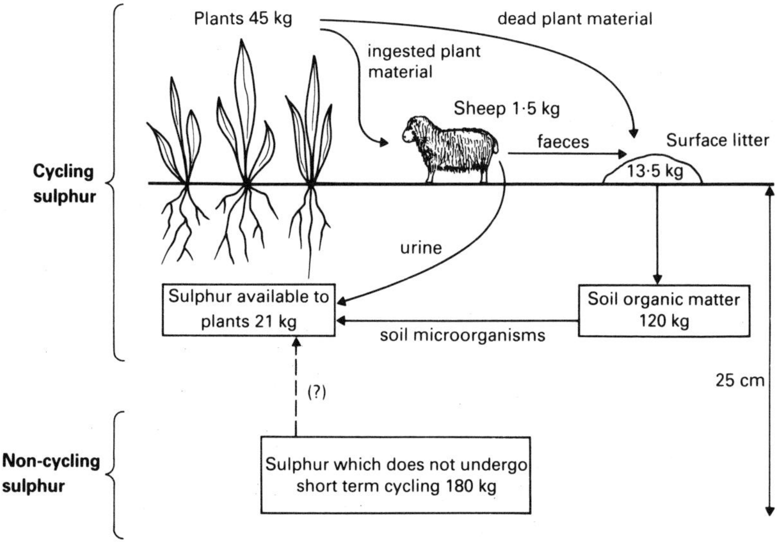

Fig. 4–5 Typical estimates of the amount of sulphur associated with some components of the soil–plant–animal system used for wool production in the New England district of N.S.W., Australia. The values are expressed as kg hectare^{-1} and apply for a stocking rate of about 20 sheep hectare^{-1} on a pasture comprised of grasses and clovers top-dressed with approximately 200 kg of superphosphate hectare^{-1} year^{-1} (200 kg superphosphate contains about 23 kg of sulphur). The estimates for the soil component apply to a depth of about 25 cm.

however, a large proportion of the cycling sulphur is tied up in organic matter, some on the soil surface in dead plant material and animal faeces, but mostly in the soil. The rate of conversion of sulphur in organic matter to forms which can be used by plants is the slowest step in sulphur cycling and therefore determines the overall rate of sulphur cycling in soils which have an inherently low sulphur content.

5 Man's Disturbance of the Sulphur Cycle and its Consequences

In Chapter 4 it was noted that the total amount of sulphur removed annually from the sulphur reserves by man was approximately 106 million tonnes. Approximately 60% of this is inadvertently released into the atmosphere as SO_2 through such processes as the burning of fossil fuels and the smelting of metal sulphides. The remainder constitutes sulphur production which is obtained either by deliberate mining or as a by-product from other commodities. Sulphur from these sources is used for the production of a large number of compounds which in turn are used in many applications. The concern in this chapter is with the effect of the various forms of sulphur produced (either intentionally or unintentionally) by man on organisms. The most important of these is the inadvertent production of atmospheric SO_2 but the intentional uses of sulphur by man must also be considered, since some of the products of these processes have important effects on some organisms and therefore the sulphur cycle itself.

5.1 Effect of atmospheric sulphur dioxide on organisms

In the absence of any industrial activity, the concentration of SO_2 in the atmosphere is typically 1 to 3 $\mu g\ m^{-3}$. Concentrations in this range are found in the remoter (non-industrialized) parts of the world today and presumably occurred over most parts of the earth's surface prior to the industrial revolution. These natural or background concentrations are casued by a combination of many processes. The most important of these is the volatilization of H_2S by organisms (e.g. processes of desulphurylation and dissimilatory sulphate reduction). The H_2S is subsequently oxidized to SO_2 by ozone in the atmosphere:

$$H_2S + O_3 \longrightarrow H_2O + SO_2$$

Various geophysical emissions associated with volcanoes, sulphur springs, etc., also add SO_2 to the atmosphere but these are relatively unimportant. Biological volatilization of sulphur amounts to approximately 88 million tonnes per year and is fairly evenly distributed over the earth's surface. By comparison, anthropogenic emissions of sulphur amount to 63 million tonnes annually, but the bulk of this is released as SO_2 in a few localized industrial regions which constitute a very small proportion of the earth's surface. Thus anthropogenic emissions can cause high localized concentrations of SO_2 and these can be exacerbated by certain atmospheric conditions. What then are the effects of the SO_2 concentrations produced by man on organisms?

5.1 *Effect on man*

Most healthy subjects exposed to SO_2 concentrations greater than 3 mg m^{-3} (i.e. 3000 $\mu g\ m^{-3}$) show a detectable physiological response. Concentrations of 15 mg m^{-3} for one hour cause choking and subjects exposed to 30 mg m^{-3} for one hour undergo severe distress. Not all subjects, however, react in this way. Approximately 1% of the population show distress at much lower concentrations (e.g. bronchial constriction at 3 mg m^{-3}) whilst those whose occupations bring them in contact with mean concentrations of 15 mg m^{-3} develop a mild resistance although they lose their capacity to smell the gas.

How then do these values compare with those found in industrial cities? One of the most interesting and tragic examples can be seen from the data compiled during the fog in London, 5–9 December 1962. The mean value for SO_2 immediately prior and after the fog was approximately 0.4 mg m^{-3}. During the period of the fog, however, a temperature inversion prevented the escape and dissipation of smoke and flue gases (containing SO_2) from chimneys. Accordingly, the concentration of SO_2 increased approximately 5-fold to 2.1 mg m^{-3}. The smoke concentration increased by a similar amount and so did the death rate; in fact the total loss of life during the period of the fog amounted to approximately 4000 more individuals than might have been predicted from the mean daily death rate had the fog not occurred. Similar observations have been made in several other industrial towns and cities. The data of course, do not prove that the elevated concentrations of SO_2 and/or smoke caused the deaths but it is clear that the variables are closely correlated. Interestingly, the maximum SO_2 concentrations recorded during the fog are not in themselves sufficient to account for the death rate. However, as stated previously, a proportion of subjects are more sensitive to SO_2 concentration than most. Presumably, it was this effect, together with enhanced smoke concentrations and extreme cold, which contributed to the loss of life. Hopefully, with the introduction of SO_2 and smoke-abatement measures these calamities are a thing of the past.

5.1.2 *Effect on lichens*

Lichens are a symbiotic association of a fungus and an alga; a specific lichen has specific fungal and algal partners. Lichens are photoautotrophs which grow on virtually inert surfaces and are therefore unable to obtain appreciable amounts of nutrient salts from the surfaces to which they are attached. Instead they derive at least some of the nutrients they require either by direct absorption of gases present in the atmosphere or by absorption of dissolved forms of these gases present in rainwater; nitrogen for example is obtained from NH_3 and sulphur from SO_2. The mechanisms for absorbing these gases were, however, evolved to take up the very low concentrations of SO_2 and NH_3 associated with unpolluted air, a process which occurred long before the industrial revolution. Presumably, it is the evolution of a gaseous nutrition mechanism which makes lichens so especially sensitive to anthropogenic emissions of SO_2 and other pollutants. This would explain the high concentrations of sulphur found in lichens growing in polluted air. Lichens growing in unpolluted air typically contain 0.2 mg sulphur g^{-1} dry mass whereas values of 1.5 mg g^{-1} have been reported from lichens collected from polluted sites. Some lichens are amongst the most sensitive organisms to SO_2; the species *Lobaria pulmonaria* and *Teloschistes flavicans* are not usually found in some regions of the U.K. which have a mean winter SO_2 concentration greater than 30 $\mu g\ m^{-3}$ (cf. natural or background values of 1 to 3 $\mu g\ m^{-3}$ for unpolluted air). Lichens are therefore of great interest as indicators of SO_2 pollution in the atmosphere. Further, various species of lichens differ in their susceptibility to SO_2 concentration. Indeed, the differences in the sensitivity of lichen species to SO_2 are so distinct that the frequency of various corticulous species has been used to map SO_2 concentrations in relation to a source of SO_2 emission. In the U.K. for example, only *Pleurococcus viridis* grows on the trunks of trees with acidic bark when the mean winter SO_2 concentration is greater than 170 $\mu g\ m^{-3}$, *Lepraria incara* occurs at 125 $\mu g\ m^{-3}$, various species of *Pertusaria* occur at approximately 50 $\mu g\ m^{-3}$ and a large number of species occur when the air is relatively unpolluted (less than 30 $\mu g\ m^{-3}$). Since the industrialized regions of the U.K. have mean winter SO_2 concentrations in excess of 150 $\mu g\ m^{-3}$ it is apparent that the lichen flora has decreased dramatically since the Industrial Revolution. This interesting topic is described in more detail in HAWKSWORTH and ROSE, 1976.

5.1.3 Effect on higher plants

Prolonged exposure of most plants to SO_2 concentrations of 1000 $\mu g\ m^{-3}$ causes yellowing and blotching of leaves and a decrease in photosynthetic rate. Most plants are therefore less sensitive to SO_2 than lichens but are more sensitive than man. There are several classic instances on record where anthropogenic emissions of SO_2 arising from the smelting of base metal sulphide ores has caused complete devastation of the surrounding vegetation without causing any gross health problems in workers in the smelters. Although the SO_2 emissions are now regulated the results of the devastation they have caused can still be seen (Fig. 5–1).

It has already been noted that lichens use the low concentrations of SO_2 associated with unpolluted air as a sulphur source. When the concentration of available sulphur in the soil is inadequate to sustain maximum plant growth it can be demonstrated that plants can also use

Fig. 5–1 Defoliation of native vegetation by SO_2 in the vicinity of a copper sulphide smelting plant at Queenstown, Tasmania. (Photo by courtesy of M. Van Steveninck, 1960.)

gaseous SO_2 as a sulphur source. This is shown by the experiment in Fig. 5–2. *Lolium perenne* (perennial ryegrass) was grown in pots of a soil which was known to contain insufficient amounts of sulphur to support optimum growth of this species. The soil in half the pots was supplemented with inorganic sulphate, and half the plants in each treatment (with and without sulphate) were supplied with air containing 50 $\mu g\ SO_2\ m^{-3}$; the remainder were supplied with air only. Plants in all treatments were enclosed in canopies to maintain the appropriate conditions in the gas phase. It can be seen that plants supplied with sulphate (with or without SO_2) grew best and that SO_2 at 50 $\mu g\ m^{-3}$ was not toxic. In the absence of sulphate, however, the plants subjected to SO_2 grew much better than those which were not supplied with either sulphate or SO_2. It follows, therefore, that SO_2 of anthropogenic origin can, in theory, enhance plant growth in regions where the soil contains insufficient sulphur to support optimum growth. Although the experiment shown in Fig. 5–2 does not rule out the possibility that SO_2 is made available to plants by microbial oxidation to sulphate in the soil, other experiments, not shown in this booklet, rule out this possibility. Plants therefore actively remove SO_2 from the atmosphere.

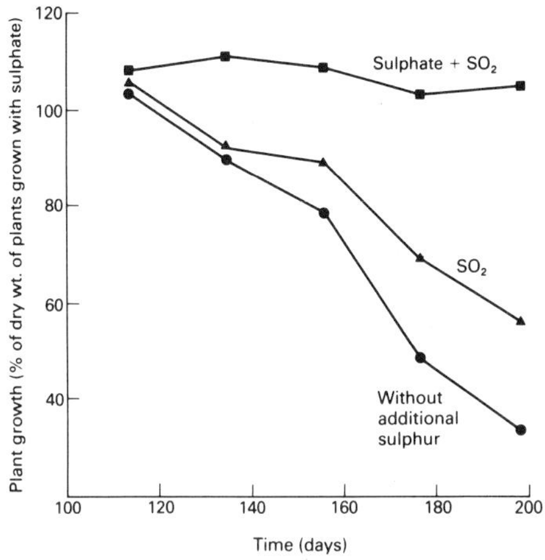

Fig. 5–2 Effect of SO_2 on growth of perennial ryegrass (*Lolium perenne*). Plants were grown in pots of soil containing insufficient sulphur for optimum growth; sulphate was added to the soil and/or SO_2 supplied in the gas phase as appropriate (see the text). The plants were harvested at intervals and allowed to regrow before being reharvested. Growth at each harvest is expressed as dry mass relative to the dry mass of plants grown in soil containing added sulphate. (Recalculated from COWLING, D. W. and LOCKYER, D. R. (1976). *J. Exp. Bot.*, **27**, 411–17.)

It is more difficult to establish that anthropogenic emissions of SO_2 have in practice enhanced plant productivity. Circumstantial evidence, however, suggests that this does occur though it is probably mediated more through the soil than by direct foliar absorption of the gas. SO_2 absorbed by the soil can be readily oxidized to sulphate by chemoautotrophic organisms so making the sulphur available to plants. Figure 5–3a illustrates the amounts of SO_2 absorbed by soil, either as gas or from rainwater, in the U.S.A. It can be seen that the accessions of sulphur are greatest in the industrial areas of the north-east and the so-called mid-west region south of the Great Lakes. Figure 5–3b shows that sulphur deficiency in these regions is uncommon in comparison with the rest of the U.S.A. Similar relationships have been reported in other countries. This suggests that the absorption of anthropogenic SO_2 emissions by soils in industrial regions might be important in preventing sulphur deficiency in crops and pastures.

Whilst it is likely that anthropogenic SO_2 emissions have increased man's food supply in certain regions, this in itself is not sufficient reason for allowing uncontrolled SO_2 emissions to continue. Quite apart from the injurious effects of the gas to plants, public health and buildings (SO_2

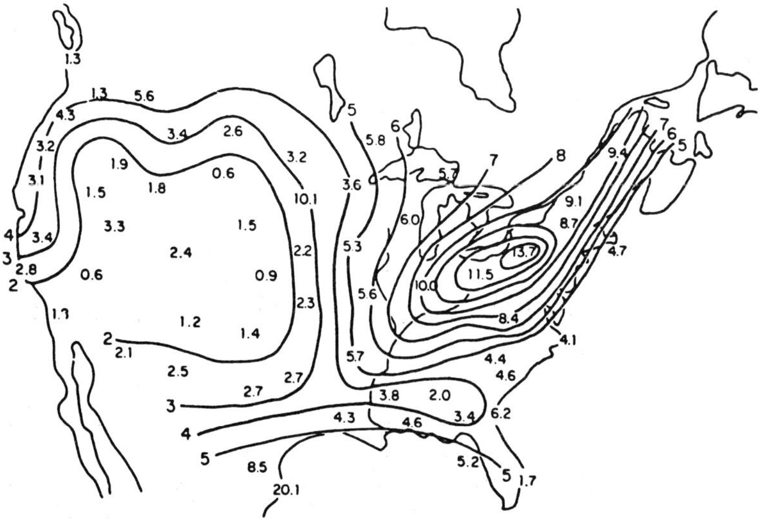

(a)

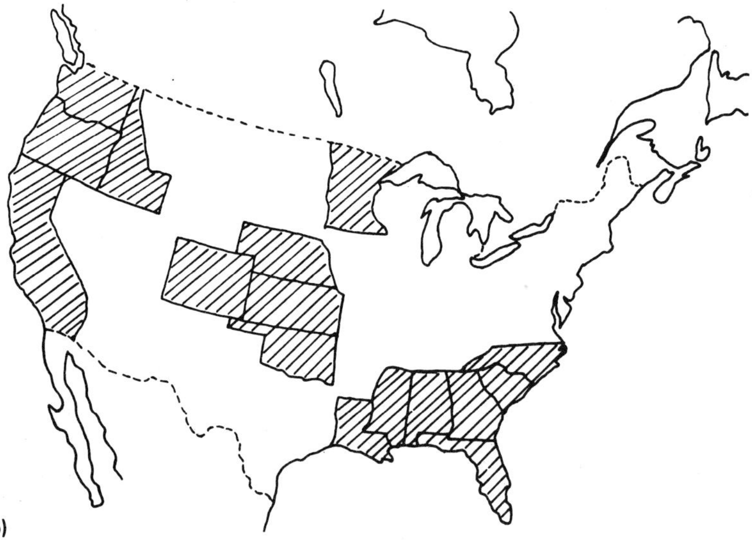

(b)

Fig. 5–3 Additions of sulphur to the soil from the atmosphere in North America (values in g m^{-2} $year^{-1}$) (**a**), and map showing the states in U.S.A. in which sulphur deficiency has been reported (**b**). (Data in (**a**) from ERIKSSON, E. (1960). *Tellus*, **12**, 63–109. (**b**) compiled from data supplied by The Sulphur Institute, Washington.)

concentration is invariably correlated with population and housing density) there are two other aspects to consider. First, most of the sulphur is added to the soil in relatively close proximity to the source of emission. It is therefore not available for use in a controlled manner in regions that are known to be sulphur deficient. Secondly, and perhaps more importantly, nothing is known about the long-term effects of absorption of low concentrations of SO_2 by the soil. Long-term additions for instance might cause a decrease in soil pH and so induce other nutritional deficiencies in plants. Continued addition of sulphur as SO_2 might also enhance eutrophication of drainage water and of the streams into which it drains (see section 5.3.2).

5.2 Uses of sulphur by man

Approximately 88% of the sulphur produced in the western world is used for the manufacture of sulphuric acid and of this, by far the greatest part is used for the manufacture of fertilizers. The two most important processes here are the manufacture of superphosphate and ammonium sulphate. Superphosphate is the product formed when $Ca_3(PO_4)_2$, an insoluble form of phosphorus known as rock phosphate, is treated with H_2SO_4 to make a soluble form of phosphorus which can be taken up by the roots of plants. It can be seen from the reaction:

$$Ca_3(PO_4)_2 + 2H_2SO_4 \longrightarrow Ca(H_2PO_4)_2 + 2CaSO_4$$

that superphosphate, which is regarded as a phosphatic fertilizer, also contains $CaSO_4$. H_2SO_4 is also used in the manufacture of the nitrogenous fertilizer ammonium sulphate to trap gaseous NH_3 produced by the Haber process or by the gasification of coal to make a salt which can be easily handled:

$$2NH_3 + H_2SO_4 \longrightarrow (NH_4)_2SO_4$$

In fact about 57% of man's annual sulphur production is used in the manufacture of fertilizers. This amount is therefore made directly available for cycling by organisms. However, very little of man's production of sulphur is intentionally used to make sulphatic fertilizers as such; nearly all of the sulphur added to soil is applied inadvertently in phosphatic and nitrogenous fertilizers in which the sulphate component can be regarded as a non-toxic contaminant resulting from the manufacturing process. Consequently, much of the sulphur applied to soil in fertilizers is, from the point of view of plant nutrition, added needlessly.

Various forms of sulphur (S^0, SO_2, H_2SO_4, etc.) are used in a large number of manufacturing processes such as the vulcanizing of rubber, pickling of steel, extraction of cellulose from wood, production of paints, explosives, rayon, detergents, and pharmaceuticals to name but a few.

Each process leads either to the manufacture of a product containing sulphur (e.g. anionic detergents) or the generation of waste products containing sulphur (e.g. sulphite waste liquors from pulping of wood). It is not possible to describe the effects of each product and by-product on organisms but one example of each type is discussed in more detail in section 5.4 as some of these can cause deleterious effects on organisms and therefore influence small scale sulphur cycling.

5·3 Effect of sulphur fertilizers on organisms

5.3.1 Effect on plant and animal productivity

Inorganic sulphate in the soil is the normal source of sulphur for the nutrition of plants. There are many regions in the world, however, where the concentration of sulphate available in the soil is too low to sustain optimum plant growth. Under these conditions, the application of a suitable sulphur fertilizer enhances plant growth and therefore animal production. In practice, a variety of compounds are used to correct sulphur deficiency. Gypsum ($CaSO_4$) is widely used and even elemental sulphur (S^0) is quite effective as the chemoautotrophs present in soil will, under aerobic conditions, rapidly oxidize S^0 to sulphate. The latter process, however, tends to decrease the pH of the soil and induce other nutritional problems. Most of the sulphur added to soils in fertilizers, however, is added unintentionally during the application of phosphatic and nitrogenous fertilizers. Nevertheless, there are many reports from various parts of the world showing that increased growth following the application of superphosphate is due to the sulphur content rather than to phosphorus; when gypsum ($CaSO_4$) is applied instead of superphosphate the same growth response is obtained. In these cases, phosphorus, a scarce commodity, has been applied unnecessarily.

Plant yield is not the only factor affected by the availability of sulphur in the soil. In section 3.5 it was noted that the sulphur supply can affect the relative amounts of sulphur-poor and sulphur-rich storage proteins in the seeds of legumes (Fig. 3–4). Similar results have been obtained for the storage proteins of some cereals. So we see that in those regions where sulphur limits plant growth the intentional and unintentional addition of sulphur to soil increases both the yield and the nutritional quality of plant material. This in turn leads to enhanced animal production.

5.3.2 Eutrophication

Fertilizers are designed to supply an element in a soluble form which is available for plant growth. One consequence of this practice is that some of the fertilizer, being highly soluble, is removed from the soil by the process of natural drainage and ends up in streams and rivers. Watercourses therefore, become enriched in soluble forms of the elements required by organisms. This leads to massive increases in the growth of autotrophs which enhance the amount of organic matter in the stream, thus causing increased growth of heterotrophic organisms which deplete the stream of oxygen. This chain of events not only kills aerobic organisms (e.g. fish), but encourages the growth of anaerobic bacteria which produce characteristic vile-smelling metabolic products, and ruin the quality of the water. The chain of events described above is referred to as eutrophication; in this case it is induced by the enrichment of drainage water with nutrients following the application of fertilizer. In general, it is the nitrogen and phosphorus content of drainage water from fertilized land which are held to be the main causative agents in the induction of eutrophication by man. However, since a large proportion of phosphatic and nitrogenous fertilizers also contain sulphur, it is inevitable that this element also contributes to eutrophication. Some of the more modern fertilizers are formulated so that the element in question is only slowly released into a soluble form for uptake by plants. This process often entails the metabolic activity of organisms present in soil. Fertilizers of this type not only ensure a more uniform and prolonged supply of the element for plants, but also decrease the severity of eutrophication.

5.4 Some examples of the effect of sulphur-containing wastes on organisms

5.4.1 *Waste liquors from wood-pulping*

The traditional technique for the large scale manufacture of wood-pulp used in the production of paper entails the digestion of wood with steam and $Mg(HSO_3)_2$ or $Ca(HSO_3)_2$ at high pressure. This procedure removes most of the lignin (a complex phenolic polymer) and various other compounds from the wood leaving the cellulose (pulp) behind. The lignin and other undesired compounds are sulphonated by the bisulphite and either released into solution or set free as a finely divided suspension which can be drained off from the cellulose. The waste material containing sulphonated lignin, residual bisulphite and other matter is known as sulphite waste liquor. This is invariably discharged into a river. Since 50% of the mass of the wood is contained in sulphite waste liquor, the amount of material discharged by a large pulping mill using the bisulphite process is enormous. When discharged into streams, the liquor is a most unpleasant pollutant. It is toxic to fish, taints and discolours drinking water and causes foaming. More importantly it has a very high biological oxygen demand, i.e. it provides substrates for the growth of heterotrophic organisms which deplete the stream of oxygen. This induces the growth of anaerobic organisms, especially dissimilatory sulphate-reducing bacteria, which thrive on the high concentrations of sulphite. The gross pollution of the Great Lakes of North America is largely attributed to discharges of sulphite waste liquor from pulping mills located on streams which flow into this body of water. The new process of thermomechanical pulping should not only decrease the pollution hazard associated with sulphite waste liquor, but also significantly decrease the enormous loss of wood associated with sulphite pulping.

5.4.2 *Sulphonated detergents in waste water*

A large number of different detergents (surface active agents) are used for industrial and domestic purposes. After use they are usually discharged in waste water where some of them present real problems as pollutants. The most troublesome of the detergents from this point of view is the group known as the alkyl benzene sulphonates which, in the mid-1960s, accounted for the bulk of detergent sales. Like all surface active agents they have a polar and non-polar group which accounts for their property to lower the opposing forces which operate at the interface between two immiscible substances. The alkyl benzene sulphonates are relatively resistant to biological decomposition. The rate of degradation is so slow that they are not removed by the traditional methods of waste water purification. They are therefore discharged into streams in treated water. Additional discharges from other towns further downstream cause the concentration to increase. The problem with the alkyl benzene sulphonates is not that they are particularly toxic in themselves, but that they decrease the rate of oxygen dissolution at the surface of the stream. They therefore tend to create conditions suitable for the growth of anaerobic organisms which, as described previously, lower the quality of the water. In the light of this experience, efforts are now being made to devise detergents which are subject to biological degradation.

5.5 Detoxification and sulphur tolerances of organisms

Some examples of pollution caused by man in his use of sulphur have been discussed. However, with the effect of man's activities in disturbing the sulphur cycle, scientists must not fall into the trap of considering only one point of view. In the case of anthropogenic SO_2 emissions, SO_2 is a pollutant, endangering public health, causing economic damage to buildings and has in the past caused gross defoliation of localized areas. On the other hand, SO_2 is absorbed by the soil where it is oxidized by chemoautotrophic bacteria to sulphate (non-toxic) which can be used for plant growth. In some regions it is likely that this process has increased plant and animal production for human consumption. Whilst the argument does not stop at this point, this example illustrates the principle that some organisms have the capacity to detoxify a pollutant made by man and feed the sulphur into the biological sulphur cycle. Other examples involving sulphur could easily be cited: H_2S for instance is far more toxic to aerobic organisms than SO_2, but is readily oxidized by photosynthetic and chemoautotrophic bacteria. Sulphite waste liquors are eventually oxidized to CO_2 and sulphate provided they are suitably diluted. The alkyl sulphate detergents (not to be confused with the alkyl benzene sulphonates) are readily subjected to microbial oxidation.

Finally, it should be noted that some organisms are tolerant of very high concentrations of various forms of sulphur. This is especially true for some of the sulphur-oxidizing bacteria. The chemoautotroph *Thiobacillus thio-oxidans*, for example, continues to oxidize sulphur to H_2SO_4 at pH 1. Some plants tolerate fairly high concentrations of sulphate. Many crucifers incorporate sulphate into non-essential sulphur compounds. Indeed, the non-essential sulphur compounds present in cabbage, broccoli, cauliflower, etc., are responsible for the pungent odour and biting flavour of these vegetables and it is common to add substantial amounts of sulphur fertilizer when growing these crops to enhance the formation of these compounds. Another group of plants which are tolerant of sulphate are the so-called gypsophilous plants. They are usually indigenous to soils which contain high concentrations of gypsum ($CaSO_4$) and are usually found in arid or semi-arid environments. However, unlike the crucifers, gypsophilous plants do not incorporate large amounts of sulphur into non-essential compounds.

It follows from the understanding of the interconversions of sulphur by organisms and the tolerances that some of them have to high concentrations of sulphur, that selected organisms can, in theory, be used as catalysts in industrial processes such as the manufacture of sulphuric acid from gypsum and the recovery of sulphur from sewage. Whether this will ever happen will depend on economic and social factors and on the availability of energy supplies in the future.

In conclusion, it can be seen from the discussion of the sulphur reserves that man is under no immediate or medium-term threat of running out of sulphur either for industry or agriculture. Indeed it is likely that SO_2 abatement measures will force man to produce more sulphur than he needs. It is apparent that man's release of sulphur into the atmosphere is not leading to any loss of sulphur. It returns to earth where it undergoes biological cycling and, at least temporarily, enhances the amount of available sulphur for plant growth. In the longer term it will slowly return to the sulphur reserves through the process of eutrophication and accumulate in muds, marshes and the floors of lakes and oceans to form the sulphur sediments of the future.

Further Reading

References marked with an asterisk contain descriptions and classifications of various microorganisms referred to in this book.

Books containing review articles on various aspects of sulphur in biology.

McLACHLAN, K. D., ed. (1975). *Sulphur in Australasian Agriculture*. Sydney University Press, Sydney.

MUTH, O. H., ed. (1970). *Symposium: Sulfur in Nutrition*. AVI Publishing Co., Westport, Connecticut.

SVENSON, B. H. and SODERLUND, G. H., eds. (1976). *Nitrogen, Phosphorus and Sulphur – Global Cycles*. SCOPE Report 7, *Ecol. Bull.*, Stockholm.

Books and articles pertaining to specific topics.

BARNES, R. A. (1977). Sulphur deposit account. *Nature*, **268**, 92–3.

GREENBERG, D. M., ed. (1975). *Metabolic pathways*, Vol. VII, *Metabolism of Sulfur Compounds*. Academic Press, New York.

LASCELLES, J. (1968). The bacterial photosynthetic apparatus. *Adv. Microbial Physiol.*, **2**, 1–42.

LE GALL, J. and POSTGATE, J. R. (1973). The physiology of sulphate-reducing bacteria. *Adv. Microbial Physiol.*, **10**, 81–133.

PARSONS, R. F. (1976). Gypsophily in plants – a review. *Am. Midl. Nat.*, **96**, 1–20.

POSTGATE,* J. R. (1968). The sulphur cycle. In: *Inorganic Sulphur Chemistry*, edited by G. N. Nickless. pp. 259–79. Elsevier, Amsterdam.

ROY, A. B. and TRUDINGER,* P. A. (1970). *The Biochemistry of Inorganic Compounds of Sulphur*. Cambridge University Press, London.

SCHIFF, J. A. and HODSON, R. C. (1973). The metabolism of sulfate. *Annu. Rev. Plant Physiol.*, **24**, 381–414.

Several other articles of interest are listed in the legends to Figures and Tables throughout this booklet. In addition the following booklets from the series 'Studies in Biology' are also relevant.

HALL, D. O. and RAO, K. K. (1977). *Photosynthesis*. Studies in Biology, no. 37. Second edition. Edward Arnold, London.

HAWKSWORTH, D. L. and ROSE, F. (1976). *Lichens as Pollution Monitors*. Studies in Biology, no. 66. Edward Arnold, London.

MELLANBY, K. (1972). *The Biology of Pollution*. Studies in Biology, no. 38. Edward Arnold, London.